Instructor's Manual
with Solutions and Software Instructions to Accompany

Modern Physics

Raymond A. Serway
James Madison University

Clement J. Moses
Utica College of Syracuse University

Curt A. Moyer
Clarkson University of Technology

SAUNDERS COLLEGE PUBLISHING

Philadelphia
New York
Chicago
San Francisco
Montreal
Toronto
London
Sydney
Tokyo

Printed in the United States of America

Serway: Instructor's Manual with Solutions and Software Instructions to accompany MODERN PHYSICS
ISBN # 0-03-004847-8

901 095 987654321

Table of Contents

Preface

This instructor's manual has been written to accompany the textbook "Modern Physics" by Raymond A. Serway, Clement J. Moses, and Curt A. Moyer. The manual is divided into two parts:

1. Solutions to the end-of-the-chapter problems found in the textbook.

2. Instructions for the software package that was written to accompany specific problems designated in the textbook as "Problems for the Computer".

Note that solutions to essentially all of the problems in the textbook are contained in this manual. We are of the strong opinion that this instructor's manual should not be made available to students. However, instructor s are free to reproduce parts of it for posting specific solutions. We welcome your comments on the solutions as presented here, and encourage you to suggest alternate solutions. We would also like to be notified of any errors you may find.

The software instructions contained in this manual were prepared to assist you in the use of a set of five programs that will run on IBM or IBM compatible machines. All five programs are contained on a single disk. To obtain a copy of this disk, contact your sales representative or write to the publisher.* Instructors using this textbook are free to make as many copies of the disk as necessary for use in their courses. Comments or suggestions regarding these programs should be sent to

Prof. Curt A. Moyer
Department of Physics
Clarkson University
Potsdam, New York 13676

Raymond A. Serway
Clement J. Moses
Curt A. Moyer
September, 1988

* **TO ORDER, contact your Saunders rep, or the Saunders regional office nearest you.**

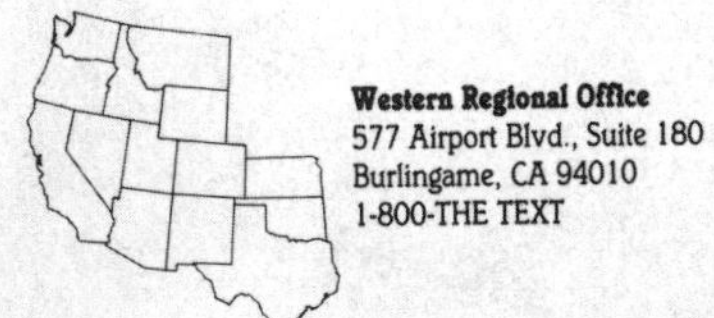

Western Regional Office
577 Airport Blvd., Suite 180
Burlingame, CA 94010
1-800-THE TEXT

Central Regional Office
901 North Elm
Hinsdale, IL 60521
1-800-227-TEXT
From Illinois, call collect
(312) 323-0205

Eastern Regional Office
744 Roble Road
Allentown, PA 18103
1-800-551-BOOK
(215) 266-1980

Canada
Holt, Rinehart and Winston
of Canada LTD.
55 Horner Ave.
Toronto, Ontario M8Z 4X6
(416) 255-4491

NS/BH. 967/88 Printed in the U.S.A.

Acknowledgements

We are most grateful to Mrs. Mary Lou Glick and Mrs. Linda Miller for their excellent work in typing this manual. We also thank Professors Don Chodrow and Joseph Rudmin for writing some of the problems in this textbook.

Part I

Solutions to Problems

Chapter 1

1. $F_1 = ma$, and $v_2 = v_1 + v_{21}$; the velocity v_{21} is the velocity of 2 with respect to 1, which is a constant. $a_2 = dv_2/dt = dv_1/dt$. Since the acceleration is measured to be the same , the force observed is also the same.

---

2. The laboratory observer notes Newton's second law to hold: $F_1 = ma$, or $m = F_1/a$, (where the subscript 1 implies the measurement was made in the laboratory frame of reference).

   The observer in the accelerating framce of reference measures the acceleration of the mass as $\mathbf{a_2 = a_1 - a'}$ (where the subscript 2 implies the measurement was made in the accelerating frame of reference, the primed acceleration term is the acceleration of the accelerated frame with respect to the laboratory frame ofreference).

   If Newton's second law held for the accelerating frame, that observer would then find valid the relation $F_2 = m/a'$ or $F_1 = m/a'$ or $m = F_1/a'$ (since $F_1 = F_{21}$ and the mass is unchanged in each). But, instead, the accelerating frame observer will find that F_2 (= F_1) = $m/(a_1 - a')$ which is <u>not</u> Newton's second law.

---

3. IN THE REST FRAME:

   $p_i = m_1v_{1i} + m_2v_{2i}$ = (2000 kg)(20 m/s) + (1500 kg)(0 m/s)
   $= 4.0 \times 10^4$ kg m/s ;
   $p_f = (m_1 = M_2)v_f$ = (2000 kg + 1500 kg)v_f ;
   $p_i = p_f \Rightarrow v_f = 4.0 \times 10^4$ kg·m/s/(2000 kg + 1500 kg) = 11.4 m/s

   IN THE MOVING FRAME:

   $v_{1i}' = v_{1i} - v'$ = 20 m/s - (- 10 m/s) = 30 m/s
   $v_{2i}' = v_{2i} - v'$ = 0 m/s - (- 10 m/s) = 10 m/s
   $p_i' = m_1v_{1i}'rm_2v_{2i}'$ = (2000 kg)(30 m/s) + (1500 kg)(10 m/s)
   $= 7.5 \times 10^4$ kg m/s

$p_f' = (2000\text{ kg} + 1500\text{ kg})v_f'$

$p_i = p_f \quad \Rightarrow v_f' = (7.5 \times 10^4\text{ kg m/s})/(2000\text{ kg} + 1500\text{ kg}) = 21.4\text{ m/s}$

$v_{1f} = v_{1f}' + v' = 21.4\text{ m/s} - 10\text{ m/s} = 11.4\text{ m/s}$

or $v_{1f}' = v_{1f} - v' = 11.4\text{ m/s} - (-10\text{ m/s}) = 21.4\text{ m/s}$ Q.E.D.

4. IN THE REST FRAME:

$p_i = m_1v_{1i} + m_2v_{2i} = (0.3\text{ kg})(5\text{ m/s}) + (0.2\text{ kg})(-3\text{ m/s}) = 0.9\text{ kg·m/s}$

$p_f = m_1v_{1f} + m_2v_{2f}$; we use Equations 9.20 and 9.21found on page 175 of the text to determine v_{1f} and v_{2f} ;

$v_{1f} = -1.4\text{ m/s}$ and $v_{2f} = 6.6\text{ m/s} \Rightarrow p_f = 0.9\text{ kg·m/s}$

IN THE MOVING FRAME:

$p_i' = m_1v_{1i}' + m_2v_{2i}'$;

where $v_{1i}' = v_{1i} - v' = 5\text{ m/s} - (-2\text{ m/s}) = 7\text{ m/s}$

Similarly $v_{2i}' = -1\text{ m/s}$ and $p_i' = 1.9\text{ kg·m/s}$.

To find p_f we use the same method as above to find $v_{1f}' = 0.6\text{ m/s}$ and $v_{2f}' = 8.6\text{ m/s}$ or use the Equations $v_{1f}' = v_{1f} - v'$ and $v_{2f}' = v_{2f} - v'$; Using these results gives $p_f' = 1.9\text{ kg·m/s}$. Q.E.D.

---

5. $T = T_0/[1 - (v^2/c^2)]^{1/2} \quad \Rightarrow v = [1 - (T_0/T)^2]^{1/2}c$;
   For $T = 2T_0 \quad \Rightarrow v = [1 - T_0/2T_0)^2]^{1/2}c = [1 - (1/4)]^{1/2}c = \underline{0.866c}$

---

6. $L = [1 - (v^2/c^2)]^{1/2}L_0$
   $\Rightarrow v = [1 - (L/L_0)^2]^{1/2}c$;
   Taking $L = L_0/2$
   $\Rightarrow v = \{1 - [(L_0/2)/L_0]^2\}^{1/2}c$
   $= \{1 - [L_0^2/2]^2\}^{1/2}c$; but $L_0 = 1$ m, therefore
   $v = (1 - 1/4)^{1/2}c = \underline{0.866c}$

---

7. $\Delta t = \gamma \Delta t' = \{1 - [v^2/(2c^2)]\}\Delta t'$

$\Rightarrow |\Delta t - \Delta t'| = v^2 \Delta t'/2c^2$

$\Rightarrow v = [2c^2(|\Delta t - \Delta t'|)/\Delta t']^{1/2}$;

Δt = (24 hrs/day)(3600 s/hr) = 86,400 s/day ;

$\Delta t'$ = (24 hrs/day)(3600 s/hr) - 1 = 86,399 s/day ;

$v = [2(|86{,}400 \text{ s/day} - 86{,}399 \text{ s/day}|)/(86{,}399 \text{ s/day})]^{1/2}c$

$= 0.0048c = \underline{1.44 \times 10^6}$ m/s

---

8. $L = L'/\gamma$; $1/\gamma = L/L' = [1 - (v/c)^2]^{1/2}$

$v = c[1 - (L/L')^2]^{1/2} = c[1 - (75/100)^2]^{1/2} = \underline{0.661}c$

---

9. $L = L'/\gamma = L'(1 - v^2/c^2)^{1/2} = L_0[1 - (0.9)^2]^{1/2} = \underline{0.436}L_0$

---

10. $T = \gamma T' = \{1 + (v^2/2c^2)\}T'$

$= \{1 + (400 \text{ m/s})^2/2(3 \times 10^8 \text{ m/s})^2\}(3600 \text{ s})$

$= \{1 + 9.0 \times 10^{-13}\}(3600 \text{ s})$

$= \underline{3600 \text{ s} + 3.2 \times 10}^{-9}$ s

---

11. a) $\tau = \gamma\tau' = \tau'/[1 - (v/c)^2]^{1/2} = (2.6 \times 10^{-8} \text{ s})/[2 - (0.95)^2]^{1/2}$

$\tau = \underline{8.33 \times 10}^{-8}$ s

b) $d = v\Delta t = (0.95)(8.33 \times 10^{-8} \text{ s}) = \underline{23.7}$ m

---

12. $u_{XA} = -u_{XB}$; $u'_{XA} = 0.7c = (u_{XA} - u_{XB})/(1 - u_{XA}u_{XB}/c^2)$

$u_{XA} = 2u_{XA}/[1 + (u_{XA}/c)^2]$ or

$u_{XA} = \underline{0.4086}c$

Chapter 1

13. $x^2 - c^2t^2 = (x')^2 - c^2(t')^2$: (1)

$x' = \{[x - vt]/[1 - (v^2/c^2)]^{1/2}\}$:

$t' = \{[t - (v/c^2)x]/[1 - (v^2/c^2)]^{1/2}\}$:

Therefore (1) becomes

$$x^2 - c^2t^2 = \{[x - vt]/[1 - (v^2/c^2)]^{1/2}\}^2 - c^2\{[t - (v/c^2)x]/[1 - (v^2/c^2)]^{1/2}\}^2$$

$$x^2 - c^2t^2 = \{[x - vt]^2 - c^2[t - (v/c^2)x]^2\}/[1 - (v^2/c^2)]$$

$$[x^2 - c^2t^2][1 - (v^2/c^2)] = x^2 - 2vxt + v^2t^2 - c^2[t^2 - 2vxt/c^2 + x^2v^2/c^4]$$

$$x^2 - x^2v^2/c^2 - c^2t^2 + v^2t^2 = x^2 - 2vxt + v^2t^2 - c^2t^2 + 2vxt - x^2v^2/c^2$$

$$\Rightarrow x^2 - x^2v^2/c^2 = x^2 - x^2y^2/c^2 \quad \text{Q.E.D.}$$

---

14. $u = (v + u')/[1 + (vu'/c^2)]$

$= (0.90c + 0.70c)/[1 + (0.90c)(0.70c)/c^2] = \underline{0.98c}$

---

15. $u'_x = (u_x - v)/(1 - u_xv/c^2)$;

$u'_x = (0.5c - 0.8c)/[1 - (0.5c)(0.8c)/c^2] = -\underline{0.5c}$

---

16. $p = m_0v/[1 - (v^2/c^2)]^{1/2}$; where $c = 3 \times 10^8$ m/s

a) $p = (1.67 \times 10^{-27}\text{ kg})(0.01c)/[1 - (0.01c/c)^2]^{1/2}$
$= \underline{5.01 \times 10}^{-21}$ kg·m/s

b) $p = (1.67 \times 10^{-27}\text{ kg})(0.5c)/[1 - (0.5c/c)^2]^{1/2}$
$= \underline{2.89 \times 10}^{-19}$ kg·m/s

c) $p = (1.67 \times 10^{-27}\text{ kg})(0.9c)/[1 - (0.9c/c)^2]^{1/2}$
$= \underline{1.03 \times 10}^{-18}$ kg·m/s

Chapter 1

17. a) $p = \gamma m_0 v = 1.90 m_0 v = m_0 v/[1 - (v/c)^2]^{1/2}$

$\Rightarrow 1.90 = 1/[1 - (v/c)^2]^{1/2}$

$\Rightarrow v = c[1 - (1/1.90)^2]^{1/2} = \underline{0.85c}$

b) No change, since rest masses cancel each other.

---

18. $E = \gamma mc^2$, $p = \gamma mu$;

$E^2 = (\gamma mc^2)^2$; $p^2 = (\gamma mu)^2$;

$E^2 - p^2c^2 = (\gamma mc^2)^2 - (\gamma mu)^2c^2 = \gamma^2\{(mc^2)^2 - (mc)^2u^2\}$

$= (mc^2)^2\{1 - u^2/c^2\}(1 - u^2/c^2)^{-1} = (mc^2)^2$ Q.E.D.

---

19. a) $E_R = m_0c^2 = (1.67 \times 10^{-27}\text{ kg})(3 \times 10^8\text{ m/s})^2 = 1.503 \times 10^{-10}\text{ J}$
$= \underline{939.4\text{ MeV}}$

b) $E = \gamma m_0c^2 = (1.503 \times 10^{-10}\text{ J})/[1 - (0.95c/c)^2]^{1/2}$

$= 4.813 \times 10^{-10}\text{ J} = \underline{3.008 \times 10^3}\text{ MeV}$

c) $K = E - m_0c^2 = 4.813 \times 10^{-10}\text{ J} - 1.503 \times 10^{-10}\text{ J}$

$= 3.31 \times 10^{-10}\text{ J} = \underline{2.069 \times 10^3}\text{ MeV}$

---

20. a) When $K = (\gamma - 1)mc^2 = 5mc^2$; $\gamma = 6$ and

$E = \gamma mc^2 = 6(0.5117\text{ MeV}) = \underline{3.07}\text{ MeV}$

b) $1/\gamma = [1 - (v/c)^2]^{1/2}$ and $v = c[1 - (1/\gamma)^2]^{1/2} = c[1 - (1/6)^2]^{1/2}$

$1/\gamma = \underline{0.986}c$

Chapter 1

21. $E = \gamma m_0c^2$; $1.5m_0c^2 = \gamma m_0c^2$

$\Rightarrow 1.5 = 1/[1 - (v^2/c^2)]^{1/2}$

$\Rightarrow v = c[1 - (1/1.5)^2]^{1/2} = \underline{0.745c}$

---

22. a) $K = 50 \times 10^9$ eV ;

$m_0c^2 = (1.67 \times 10^{-27}\ \text{kg})(3 \times 10^8\ \text{m/s})^2/(1.6 \times 10^{-19}\ \text{J/eV})$

$= 939.38$ MeV

$E = K + m_0c^2 = (50 \times 10^9\ \text{eV}) + (939.38 \times 10^6\ \text{eV}) = 50{,}939.4$ MeV

$E^2 = p^2c^2 + m_0{}^2c^4 \quad \Rightarrow p = \{[E^2 - (m_0c^2)^2]/c^2\}^{1/2}$

$p = [(50{,}939.4\ \text{MeV})^2 - (939.38\ \text{MeV})^2]^{1/2}/c = 5.04 \times 10^{10}$ eV/c; or

$p = [(5.09 \times 10^{10}\ \text{eV})/(3 \times 10^8\ \text{m/s})](1.6 \times 10^{-19}\ \text{J/eV})$
$= \underline{2.71 \times 10^{-17}}$ kg·m/s

b) $E = \gamma m_0c^2 = m_0c^2/[1 - (v/c)^2]^{1/2}$

$\Rightarrow v = c[1 - (m_0c^2/E)^2]^{1/2}$

$= (3 \times 10^8\ \text{m/s})[1 - (939.38\ \text{MeV}/50{,}939.4\ \text{MeV})^2]^{1/2}$

$= \underline{2.9995 \times 10}^8$ m/s

---

23. $\Delta E = (\gamma_1 - \gamma_2)mc^2$ and for an electron $mc^2 = 0.5117$ MeV

a) $\Delta E = \{1/[1 - (0.75)^2]^{1/2} - 1/[1 - (0.5)^2]^{1/2}\}0.5117$ MeV $= \underline{0.183}$ MeV

b) $\Delta E = \{1/[1 - (0.99)^2]^{1/2} - 1/[1 - (0.9)^2]^{1/2}\}0.5117$ MeV $= \underline{2.45}$ MeV

---

24. $\Delta m = m_{Ra} - m_{Rn} - m_{He}$

$\Delta m = (226.0254 - 222.0175 - 4.0026)$ u

$\Delta m = (\Delta m)(931 \text{ MeV/u}) = (0.0053 \text{ u})(931 \text{ MeV/u}) = \underline{4.98}$ MeV

---

25. (a) $\Delta m = 54.9279 \text{ u} - 54.9244 \text{ u} = 0.0035 \text{ u}$

$\Delta E = (939 \text{ MeV/u})(0.0035 \text{ u}) = \underline{3.29 \text{ MeV}}$

(b) The rest energy for an electron is 0.512 MeV. Therefore,

$K = 3.29 \text{ MeV} - 0.512 \text{ MeV} = \underline{2.77 \text{ MeV}}$

---

26. (a) $\Delta m = 6m_p + 6m_n - m_C = 6(1.007825) + 6(1.008665) - 12$
$= \underline{0.09894 \text{ u}}$

(b) $\Delta E = (939 \text{ MeV/u})(0.09894 \text{ u}) = 92.9$ MeV

Therefore the energy per nucleon = $92.3/12 = \underline{7.74 \text{ MeV}}$

---

28. (a) $\tau = \gamma\tau' = [1 - (0.95)^2]^{-1/2}(2.2\ \mu\text{s}) = \underline{7.05}\ \mu$s

(b) $\Delta t' = d/0.95c = (3 \times 10^{10} \text{ m})/(0.95c \text{ m/s}) = \underline{10.5}\ \mu$s

Therefore $N = (5 \times 10^4 \text{ muons}) \exp[-(1.05 \times 10^{-5} \text{ s})/(2.2 \times 10^{-6} \text{ s})]$
$= \underline{423}$ muons

---

29. a) When $K_e = K_p$, $m_e c^2(\gamma_e - 1) = m_p c^2(\gamma_p - 1)$

In this case $m_e c^2 = 0.5117$ MeV and $m_p c^2 = 939$ MeV

$\gamma_e = [1 - (0.75)^2]^{-1/2} = 1.5119$

Substituting these values into the first equation, we find

$\gamma_p = m_e c^2(\gamma_e - 1)/(m_p c^2) + 1 = 0.5117(1 - 1.5119)/937 + 1$
$= 1.000279$

But $\gamma_p = 1/[1 - (u_p/c)^2]^{1/2}$; therefore
$u_p = c(1 - \gamma_p^{-2})^{1/2} = \underline{0.0236}c$

b) Using Eq. 1.21 we have when $p_e = p_p$; $\gamma_p m_p u_p = \gamma_e m_e u_e$ or

$u_p = (\gamma_e/\gamma_p)(m_e/m_p)u_e$

$u_p = (1.5119/1.000279)[(0.5117/c^2)/(939/c^2)](0.75c)$

$u_p = \underline{6.18 \times 10^{-4}}\ c$

---

31. (a) $\ell_0 = [(\ell_x')^2 + (\ell_y')^2]^{1/2} = [\ell_0^2\cos^2\theta + \ell_0^2\sin^2\theta]^{1/2}$ so that,

$\ell = [(\ell_0^2\cos^2\theta/\gamma^2) + \ell_0^2\sin^2\theta]^{1/2} = \ell_0[1 - (v^2/c^2)\cos^2\theta]^{1/2}$

(b) $\tan\theta_0 = \ell_y'/\ell_x' = \ell_y/\gamma\ell_x = (1/\gamma)\tan\theta$

---

32. (a) For edge parallel to direction of motion, ℓ is observed to have a length ℓ/γ. The other two edge lengths are observed unchanged; therefore

$$V' = (\ell)(\ell)(\ell/\gamma) = \ell^3/\gamma$$

b) $\rho' = m'/V' = \gamma m/(\ell^3/\gamma) = \gamma^2\rho$

---

33. a) The speed as observed in the laboratory is found by using Eq.1.20:

$u_x = (u_x' + v)/(1 + u_x'v/c^2)$

But $u_x' = c/n$ (speed measured by an observer moving with the fluid.)

Therefore $u_x = [(c/n) + v]/[1 + v/(nc)] = (c/n)(1 + nv/c)/[1 + v/(nc)]$

b) $v/c << 1$, $u_x' = (c/n)[1 + n(v/c)][1 - v/(nc)]$

$$u_x' = (c/n)[1 + nv/c - v/(nc) - v^2/c^2] = c/n + v - v/n^2$$

---

34. a) $\Delta t = \gamma \Delta t'$; $\Delta t = 1/f$; $\Delta t' = 1/f_0$

Therefore, $f = (1/\gamma) f_0 = [1 - v^2/c^2]^{1/2} f_0$

b) If $v/c << 1$, $\gamma = 1/[1 - v^2/c^2]^{1/2} \approx 1 + v^2/2c^2$ and

$$\Delta f/f = f_0(1/\gamma - 1)/f_0(1/\gamma) = v^2/2c^2$$

$$\Delta f/f_0 = f_0(1/\gamma - 1)/f_0 = - v^2/2c^2$$

c) $v^2 = 3kT/m = 3(1.38 \times 10^{-23} \text{ J/k})(300 \text{ K})/(1.67 \times 10^{-27} \text{ kg})$

$$v^2 = 7.437 \times 10^6 \text{ m}^2/\text{s}^2$$

Using the result from part b, we have

$$|\Delta f/f_0| = (7.437 \times 10^6)^2/2c^2 = \underline{4.13 \times 10^{-11}}$$

---

35. $f_0 = 1/T_0 = (c - v)/\lambda_0 = (c - v)/(\lambda/\gamma) = (c - v)\gamma/(c/f)$

$f_0 = [(c - v)/c][1/(1 - v^2/c^2)]^{1/2} f$

Therefore $f_0 = [(c - v)/(c + v)]^{1/2} f$.

If the source changes direction, $v \to - v$, and

$$f_0 = [(c + v)/(c - v)]^{1/2} f$$

---

36. (a) $f_{oc} = [(c - v_s)/(c + v_s)]^{1/2} f$; and when $v_s << c$

$$[(c - v_s)/(c + v_s)]^{1/2} \approx (1 - v_s/2c)(1 - v_s/2c) \approx 1 - v_s/c$$

Therefore $\gamma_0 = (1 - v_s/c)\lambda$ or $(\lambda - \lambda_c)/\lambda \approx v_s/c$

(b) $v_s = c(200/3970) = \underline{0.0504}c$

Chapter 2

1. Using $E = hf$, with $h = 4.136 \times 10^{-15}$ eV·s gives

   (a) for $f = 5 \times 10^{14}$ Hz, $E = \underline{2.07 \text{ eV}}$

   (b) for $f = 10$ GHz, $E = \underline{4.14 \times 10^{-5}}$ eV

   (c) for $f = 30$ MHz, $E = \underline{1.24 \times 10^{-7}}$ eV.

---

2. Use $E = hc/\lambda$, or $\lambda = hc/E$ (where $hc = 1242$ eV·nm) and the results ot problem 1 to find

   (a) $\lambda = \underline{600 \text{ nm}}$, (b) $\lambda = \underline{0.03 \text{ m}}$ and (c) $\lambda = \underline{10 \text{ m}}$

---

3. The energy per photon = hf, and the total energy E transmitted in a time t is Pt, where $P = 100$ kW. Since $E = Nhf$, where N is the total number of photons transmitted in the time t, and $f = 94$ MHz, we have

$$Nhf = (100 \text{ kW})t = (10^5 \text{ W})t,$$

or

$$N/t = (10^5 \text{ W})/hf = (10^5 \text{ J/s})/(6.62 \times 10^{-34} \text{ J·s} \times 94 \times 10^6 \text{ s}^{-1})$$

$$= \underline{1.61 \times 10^{30}} \text{ photons/s}$$

---

4. Following the same reasoning as in Problem 3, we have

$$N/t = P/hf = P\lambda/hc$$

$$= (3.74 \times 10^{26} \text{ J/s})(500 \times 10^{-9} \text{ m})/(6.62 \times 10^{-34} \text{ J·s})(3 \times 10^8 \text{ m/s})$$

$$= \underline{9.44 \times 10^{44}} \text{ photons/s}$$

Chapter 2

5. Use Wien's displacement law, $\lambda_{max}T = 0.2898 \times 10^{-2}$ m·K, with $T = 35^0C = 308$ K to find

$$\lambda_{max} = (0.2898 \times 10^{-2} \text{ m·K})/(308 \text{ K}) = 9.41 \times 10^{-6} \text{ m} = \underline{9410 \text{ nm}}$$

---

8. (a) $K = hf - \phi = hc/\lambda - \phi$

$$K = (1242 \text{ eV·nm})/(350 \text{ nm}) - 2.24 \text{ eV} = \underline{1.31} \text{ eV}$$

(b) At $\lambda = \lambda_c$, $K = 0$ and $\lambda = hc/\phi$

$$\lambda = (1242 \text{ eV·nm})/(2.24 \text{ eV}) = \underline{554} \text{ nm}$$

---

10. (a) $\phi = hc/\lambda - K$

$$\phi = (1242 \text{ eV·nm})/(300 \text{ nm}) - 2.23 \text{ eV} = \underline{1.91} \text{ eV}$$

(b) $eV_s = hc/\lambda - \phi$

$$V_s = (1242 \text{ eV.nm})/(400 \text{ nm}) - 1.91 \text{ eV} = \underline{1.20} \text{ eV}$$

---

11. The energy of one photon of light of wavelength 300 nm is

$$E = hc/\lambda = (1242 \text{ eV·nm})/(300 \text{ nm}) = 4.14 \text{ eV}$$

(a) Since lithium and iron have work funtions that are less than 4.14 eV, they will exhibit the photoelectric effect for incident light with this energy. However, mercury will not since it's work function is greater than 4.14 eV.

(b) The maximum kinetic energy is given by $K = hc/\lambda - \phi$, so

$$K(\text{Li}) = (1242 \text{ eV·nm})/(300 \text{ nm}) - 2.3 \text{ eV} = \underline{1.84} \text{ eV}$$

$$K(\text{Fe}) = (1242 \text{ eV·nm})/(300 \text{ nm}) - 3.9 \text{ eV} = \underline{0.24} \text{ eV}$$

---

12. (a) $K_{max} = eV_s = e(0.45\ V) = \underline{0.45}\ eV$

(b) $\phi = hc/\lambda - K = (1242\ eV{\cdot}nm)/(500\ nm) - 0.45\ eV = \underline{2.03}\ eV$

(c) $\lambda_c = hc/\phi = (1242\ eV{\cdot}nm)/(2.03\ eV) = \underline{612}\ nm$

---

13. $\phi = 2\ eV$; $K_{max} = eV_0 = hf - \phi = (nc/\lambda) - \phi$

$\Rightarrow V_0 = [(hc/\lambda) - \phi]/e$

$= \{[(4.14 \times 10^{-15}\ eV{\cdot}s)(3 \times 10^8\ m/s)/(350 \times 10^{-9}\ m)] - 2\ eV\}/e$

$= \underline{1.55\ V}$

---

14. $K_{max} = hf - \phi = (hc/\lambda) - \phi$ $\Rightarrow \phi = (hc/\lambda) - K_{max}$;

FIRST SOURCE: $\phi = hc/\lambda - 1\ eV$

SECOND SOURCE: $\phi = hc/(\lambda/2) - 4\ eV = 2hc/\lambda - 4\ eV$

Since the work function is the same for both sources (a property of the metal), we have

$hc/\lambda - 1\ eV = 2hc/\lambda - 4\ eV$

$\Rightarrow 0 = -hc/\lambda + 3\ eV$ $\Rightarrow hc/\lambda = 3\ eV$ and

$\phi = hc/\lambda - 1\ eV = 3\ eV - 1\ eV = \underline{2 eV}$

---

15. $E = hc/\lambda$

$= (6.626 \times 10^{-34}\ J\cdot s)(3 \times 10^{8}\ m/s)/(5 \times 10^{-7}\ m)(1.6 \times 10^{-19}\ J/eV)$

$E = \underline{2.48}\ eV$

$p = h/\lambda = E/c = (2.48\ eV)(1.6 \times 10^{-19}\ J/eV)/(3 \times 10^{8}\ m/s)$

$p = \underline{1.33 \times 10}^{-27}\ kg\cdot m/s$

---

16. (a) $\Delta\lambda = h/mc(1 - \cos\theta)$

$\Delta\lambda = [6.625 \times 10^{-34}\ J\cdot s/(9.1 \times 10^{-31}\ kg)(3 \times 10^{8}\ m/s)](1 - \cos\theta)$

$\Delta\lambda = (2.43 \times 10^{-12}\ m)(1 - \cos\theta)$;

When $\theta = 90°$, $\Delta\lambda = \underline{0.00243}\ nm$

(b) Conservation of energy requires that

$hc/\lambda_0 = hc/\lambda + K_e$ or $K_e = hc(1/\lambda_0 - 1/\lambda)$

$K_e = [(6.625 \times 10^{-34}\ J\cdot s)(3 \times 10^{8}\ m/s)/(1.6 \times 10^{-19}\ J/eV)]$
$\times [(2 \times 10^{-10}\ m)^{-1} - (2.0243 \times 10^{-10}\ m)^{-1}]$

$K_e = \underline{74.5}\ eV.$

---

17. $E = 300\ keV$, $\theta = 30°$

(a) $\Delta\lambda = \lambda - \lambda_0 = h(1 - \cos\theta)/m_0c$

$= (6.63 \times 10^{-34}\ J\cdot s)(1 - \cos 30°)/[(9.11 \times 10^{-31}\ kg)(3 \times 10^{8}\ m/s)]$

$= 3.25 \times 10^{-13}\ m = \underline{3.25 \times 10}^{-4}\ nm$

17. (b) $eV = hc/\lambda$

$$\Rightarrow \quad \lambda = hc/eV = (4.14 \times 10^{-15}\ eV\cdot s)(3 \times 10^{8}\ m/s)/(300 \times 10^{3}\ eV) = 4.14 \times 10^{-12}\ m\ ;$$

So $\lambda' = \lambda + \Delta\lambda = 4.14 \times 10^{-12}\ m + 0.325 \times 10^{-12}\ m = 4.465 \times 10^{-12}\ m$

and $eV' = hc/\lambda' \Rightarrow V' = hc/\lambda' e$

$$\Rightarrow V' = [(4.14 \times 10^{-15}\ eV\cdot s)(3 \times 10^{8}\ m/s)]/[e(4.465 \times 10^{-12}\ m)] = \underline{2.78 \times 10^{5}\ eV}$$

(c) $(hc/\lambda_0) = (hc/\lambda) + K_e$

$$\Rightarrow K_e = hc[(1/\lambda_0) - (1/\lambda)]$$

$$= (4.14\times 10^{-15}\ eV\cdot s)(3\times 10^{8}\ m/s)[(1/4.14 \times 10^{-12}) - (1/4.465 \times 10^{-12})]$$

$K_e = \underline{22\ keV}$

---

18. (a) $\Delta\lambda = h(1 - \cos\theta)/m_0c$

$$= (6.63 \times 10^{-34}\ J\cdot s)(1 - \cos\theta)/[(9.11 \times 10^{-31}\ kg)(3 \times 10^{8}\ m/s)]$$

$$= (2.426 \times 10^{-12}\ m)(1 - \cos\theta)$$

For $\theta = 30°$:

$$\Delta\lambda = (2.426 \times 10^{-12}\ m)(1 - \cos 30°) = 3.25 \times 10^{-13}\ m;$$

$$\lambda' = \lambda + \Delta\lambda,$$

$$\lambda' = 0.04 \times 10^{-9}\ m + 3.25 \times 10^{-13}\ m = \underline{4.03 \times 10^{-11}}\ m\ ;$$

18. (b) $(hc/\lambda) = (hc/\lambda') + K_e$

$$\Rightarrow \quad K_e = hc[(1/\lambda) - (1/\lambda')]$$

For $\theta = 30°$:

$$K_e = (6.63 \times 10^{-34}\,J{\cdot}s)(3 \times 10^8\,m/s)[(1/0.04 \times 10^{-9}) - (1/4.03 \times 10^{-11})]$$

$$= (3.70 \times 10^{-17}\,J)[1/(1.6 \times 10^{-19}\,J/eV) = \underline{231\ eV}\ ;$$

The rest of the calculations are similar and the following table summarizes the values.

$\theta°$	$\Delta\lambda$(nm)	λ'(nm)	E(ev)
30	0.00325	0.0403	231
60	0.00121	0.0412	905
90	0.00243	0.0424	1759
120	0.00364	0.0436	2566
150	0.00453	0.0445	3142
180	0.00485	0.0448	3330

(c) The electron which is backscattered, corresponding to $\theta = 180°$.

---

19. $V_s = (h/e)f - \phi/e$. Choosing two points on the graph we have

$$0 = (h/e)(4 \times 10^{14}\,Hz) - \phi/e \quad \text{and} \quad 1.7\,V = (h/e)(8 \times 10^{14}\,Hz) - \phi/e$$

Combining these two expressions we find:

(a) $\phi = \underline{1.7}$ eV

(b) $h/e = \underline{4.0 \times 10^{-15}}\ V{\cdot}s$

(c) For cutoff wavelength, $\lambda_c = hc/\phi = (1242\ eV{\cdot}nm)/(1.7\ eV)$

$\lambda_c = \underline{730}$ nm

---

20. (a) The total energy of a simple harmonic oscillator having an amplitude A is $kA^2/2$, therefore

$$E = kA^2/2 = (25\ \text{N/m})(0.4\ \text{m})^2/2 = \underline{2.0}\ \text{J}$$

The frequency of oscillation is

$$f = (1/2\pi)(k/m)^{1/2} = (1/2\pi)(25/2)^{1/2} = \underline{0.56\ }\text{Hz}$$

(b) If the energy is quantized, we have $E_n = nhf$, and from the result from part (a) we have

$$E_n = nhf = n(6.63 \times 10^{-34}\ \text{J·s})(0.56\ \text{Hz}) = 2.0\ \text{J}$$

Therefore, $n = \underline{5.4 \times 10^{33}}$

(c) The energy carried away in a one-quantum change of energy is

$$E = hf = (6.63 \times 10^{-34}\ \text{J·s})(0.56\ \text{Hz}) = \underline{3.7 \times 10^{-34}}\ \text{J}$$

---

21. The force acting on a charge moving perpendicular to a magnetic field has a magnitude given by qvB. The charge moves in a circle of radius R, and from Newton's second law we have $F = qvB = mv^2/r$, or $v = qBr/m$. Hence, we can express the kinetic energy of the charge as

$$K = mv^2/2 = (qBr)^2/2m.$$

Using the photoelectric equation $K = hc/\lambda - \phi$, we get

$$\phi = hc/\lambda - (qBr)^2/2m$$

Substituting in the values $hc = 1242$ eV·nm, $\lambda = 450$ nm, $B = 2 \times 10^{-5}$ T, $r = 0.2$ m, $q = 1.6 \times 10^{-19}$ C, and $m = 9.11 \times 10^{-31}$ kg, gives

$$\phi = 2.76\ \text{eV} - 1.41\ \text{eV} = \underline{1.35\ \text{eV}}$$

---

22. **Symmetric Scattering, $\theta = \phi$:** First, use the equations of conservation of momentum given by Equations 2.30 and 2.31 for this two-dimensional scattering process, with $\theta = \phi$:

$$hc/\lambda_o = (hc/\lambda')\cos\theta + p_e\cos\theta \qquad (1)$$

$$(hc/\lambda')\sin\theta = p_e\sin\theta$$

or

$$p_e = hc/\lambda' \qquad (2)$$

Substitute (2) into (1) to give

$$\lambda' = 2\lambda_o\cos\theta \qquad (3)$$

Next, let us express the Compton scattering formula as

$$\lambda' - \lambda_o = \lambda_c(1 - \cos\theta) \qquad (4)$$

where $\lambda_c = h/mc = 0.00243$ nm. Combining (4) and (3) gives

$$\cos\theta = (\lambda_c + \lambda_o)/(\lambda_c + 2\lambda_o)$$

In this case, since E = 1.02 MeV, and $E = hc/\lambda_o$, we have

$$\lambda_o = hc/E = (1242 \text{ eV·nm})/(1.20 \times 10^6 \text{ eV}) = 0.00122 \text{ nm}$$

Therefore,

$$\cos\theta = (0.00243 + 0.00122)/(0.00243 + 0.00244) = 0.7495$$

$$\theta = \underline{41.5^0}$$

(b) $\lambda' = \lambda_o + \lambda_c(1 - \cos\theta)$

$$\lambda' = 0.00122 \text{ nm} + (0.00243 \text{ nm})(1 - \cos 41.5^0) = 0.00183 \text{ nm}$$

$$E = hc/\lambda' = (1242 \text{ eV·nm})/(0.00183 \text{ nm}) = \underline{0.679 \text{ MeV}}$$

23. $(hc/\lambda) = (hc/\lambda') + K_e \quad \Rightarrow K_e = hc[(1/\lambda) - (1/\lambda')] = hc\,\Delta\lambda/\lambda\lambda'$

The maximum energy is transferred to the electron when $\theta = 180°$:

$$\Delta\lambda = \lambda' - \lambda_o = \lambda_c(1 - \cos 180) = 2\lambda_c = 0.00486 \text{ nm}$$

$K_e = 30 \text{ keV} \approx (1240 \text{ eV·nm})\,\Delta\lambda/\lambda^2$

$\lambda^2 = (1240 \text{ eV·nm})(0.00486 \text{ nm})/(30 \times 10^3 \text{ eV}$

$\lambda = \underline{0.0142 \text{ nm}}$

---

26. $\lambda_c = h/mc \quad$ and $\quad \lambda = h/p$:

$\Rightarrow \lambda_c/\lambda = (h/mc)/(h/p) = p/mc$;

$E^2 = c^2p^2 + (mc^2)^2$

$\Rightarrow p = [(E^2/c^2) - (mc)^2]^{1/2}$

$\Rightarrow \lambda_c/\lambda = [(E^2/c^2) - (mc)^2]^{1/2}/mc = \{[(E^2/c^2) - (mc)^2]/(mc)^2\}^{1/2}$

$= \{[E/(mc^2)]^2 - 1\}^{1/2}$

---

27. a) Using Eq. 2.9, $\quad u(\lambda,T) = 2\pi hc^2/\{\lambda^5[e^{hc/\lambda kT} - 1]\}\quad$ we have

$$\int_0^\infty u(\lambda,T)d\lambda = \int_0^\infty 2\pi hc^2 d\lambda/\{\lambda^5[e^{hc/\lambda kT} - 1]\}$$

Change variables by letting $\quad x = hc/(\lambda kT) \quad$ and $\quad dx = -hc d\lambda/(kT\lambda^2)$

Then $$\int_0^\infty u(\lambda,T)d\lambda = -[2\pi k^4T^4/(h^3c^2)] \int_0^\infty x^3dx/(e^x - 1)$$

$= [2\pi k^4T^4/(h^3c^2)](4\pi/15)$

(Note that as λ varies from $0 \to \infty$, x varies from $\infty \to 0$)

$$\text{Therefore} \quad \int_0^\infty u(\lambda,T)d\lambda = [2\pi^5k^4/(15\,h^3c^2)]T^4 = \sigma T^4$$

(b) From part (a), $\sigma = 2\pi^5k^4/(15h^3c^2)$

$\sigma = 2\pi^5(1.38 \times 10^{-23}\ \text{J/°K})^4/[(15)(6.626 \times 10^{-34}\ \text{J·s})^3(3 \times 10^8\ \text{m/s})^2]$

$\sigma = \underline{5.7 \times 10}^{-8}\ \text{W/m}^2\text{·K}$

---

28. (a) $P/A = \sigma T^4 = (5.7 \times 10^{-8}\ \text{W/m}^2\text{·K})(3000\ \text{K}) = \underline{4.62}\ \text{MW/m}^2$

(b) $A = P/(4.62 \times 10^6\ \text{W}) = [75/(4.62 \times 10^6)]\text{m}^2 = \underline{16.2}\ \text{mm}^2$

---

29. (a) $\partial U/\partial\lambda = 0$

$$= 8\pi hc\{e^{hc/3kT}[hc/\lambda kT - 5] + 5\}/[\lambda^6(e^{hc/\lambda kT} - 1)^2]$$

or $\quad e^{hc/\lambda kT} = 1/[1 - hc/(5\lambda kT)]$

Since this condition holds for any λ, and $T > 0$ we have

$\lambda T =$ constant.

(b) From the above: $hc/\lambda kT = 5[1 - e^{-hc/\lambda kT}]$

Solve by recursion: $hc/\lambda kT = 4.965$ and

$\lambda T = \underline{2.90 \times 10}^{-3}\ \text{m·deg}$

---

32. (a) $E \cdot 2\pi r = \pi r^2 (dB/dt)$

$$E = (r/2)(dB/dt)$$

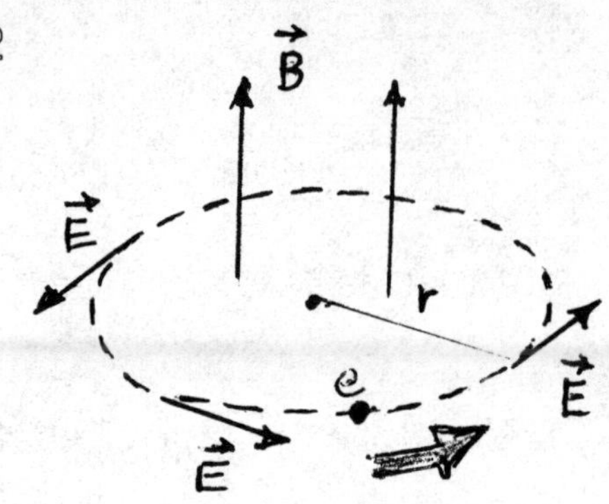

(b) $F = Eq = (r/2)(dB/dt) \cdot e$,

so $F\,dt = (re/2)(dB/dt) \cdot dt = m_e dv$

or $dv = (re/2m_e)dB$

$$\int_{v}^{v+\Delta v} dv = (er/2m_e)\int_{0}^{B} dB$$

$$\Delta v = erB/2m_e$$

(c) $\Delta\omega = \Delta v/r = eB/2m_e$

$\Delta\omega = (1.6 \times 10^{-19}\text{ C})(1\text{ T})/(2)(9.1 \times 10^{-31}\text{ kg}) = \underline{0.088 \times 10^{12}}\text{ Hz}$

(d) $\Phi_M = 0$

So $E = 0$ and there is no force on the electron.

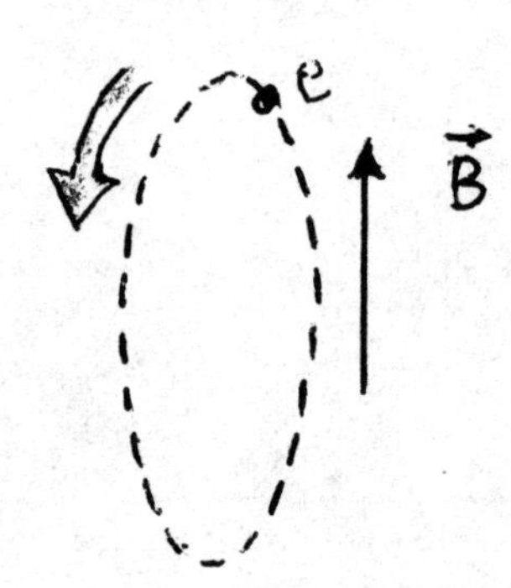

36. (a)

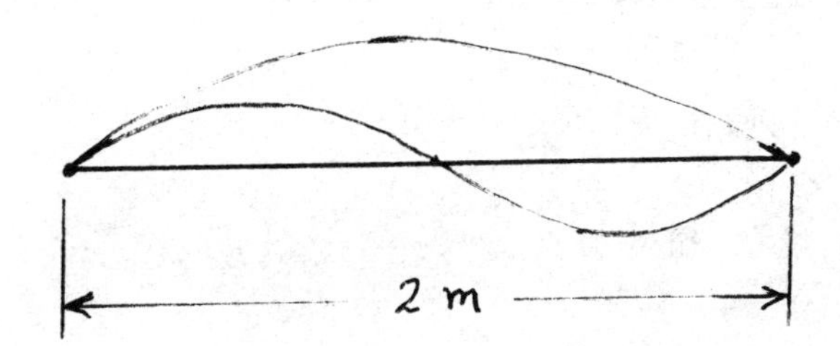

In general $n\lambda/2 = L$ where n = 1, 2, 3, . . . defines a mode or standing wave pattern with a given wavelength. Since we want to find the number of possible n-values between 2.0 and 2.1 cm, we use

$$n = 2L/\lambda$$

$$n(2.0\text{ cm}) = (2)(200)/(2.0) = 200$$
$$n(2.1\text{ cm}) = (2)(200)/(2.1) = 190.4 \approx 190$$

$$|\Delta n| = 200 - 190 = 10$$

Since n changes by one for each allowed standing wave, there are 10 standing waves of different wavelength between 2.0 and 2.1 cm.

(b) The number of modes per unit wavelength per unit length is

$$\Delta n/(\Delta\lambda L) = 10/(0.1)(200) = \underline{0.5\text{ cm}^{-2}}$$

(c) For short wavelengths n is almost a continuous function of λ. Thus we may use calculus to approximate

$\Delta n/(\Delta\lambda L) = (1/L)(dn/d\lambda)$:   Since $n = 2L/\lambda$,

$|dn/d\lambda| = 2L/\lambda^2$ and $(1/L)(dn/d\lambda) = 2/\lambda^2$

This gives approximately the same result as found in (a):

$$(1/L)(dn/d\lambda) = 2/\lambda^2 = 2/(2.0\text{ cm})^2 = \underline{0.5\text{ cm}^{-2}}$$

(d) See part (c).

38.

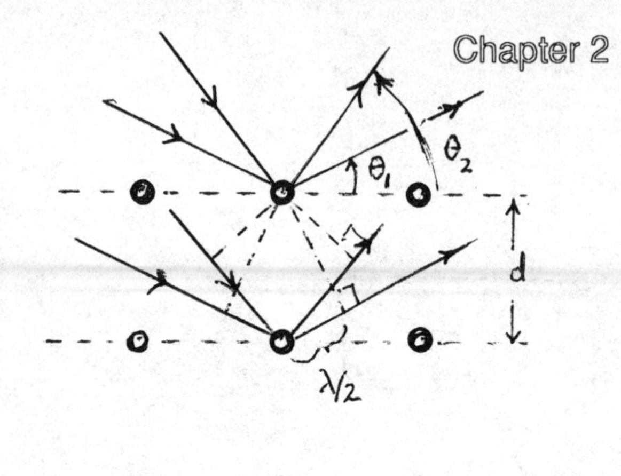

$1\lambda = 2d\sin\theta_1$ or

$\sin\theta_1 = \lambda/2d = (0.626 \times 10^{-10}\text{ m})/(8 \times 10^{-10}\text{ m})$

$\theta_1 = 0.0783$ radians = $\underline{4.49^0}$

$2\lambda = 2d\sin\theta_2$ or

$\sin\theta_2 = \lambda/d = (0.626 \times 10^{-10}\text{ m})/(4.0 \times 10^{-10}\text{ m}) = 0.1565$

$\theta_2 = \underline{9.00^0}$

$3\lambda = 2d\sin\theta_3$ or

$\sin\theta_3 = 3\lambda/2d = (3)(0.626 \times 10^{-10}\text{m})/(8.0 \times 10^{-10}\text{ m}) = 0.23475$

$\theta_3 = \underline{13.6^0}$

39.

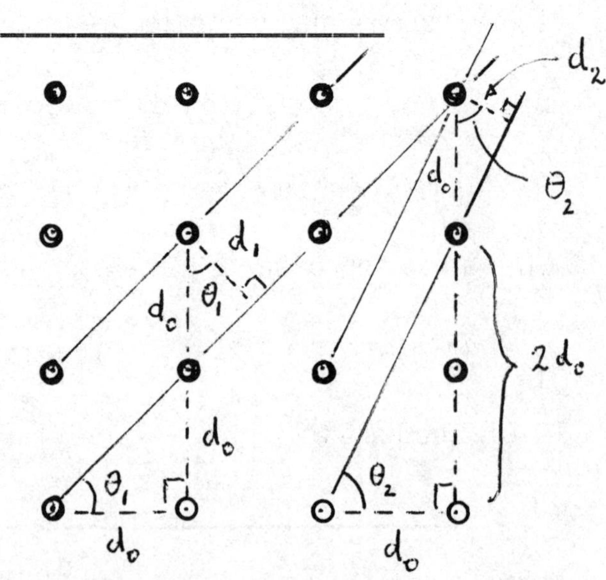

(a) $\theta_1 = 45^o$

Thus $d_1 = d_o\cos 45^o = d_o/\sqrt{2}$

$d_2 = d_o\cos\theta_2 = d_o/\sqrt{5}$

(b) The angles with respect to the d_1 planes are given by

$2d_1\sin\theta = n\lambda$ or

$\sin\theta = (1)\lambda/2d_1 = (0.626\ A)/(2)(2.83\ A) = 0.111$ $\theta = 6.37^0$

$\sin\theta = 2\lambda/2d_1 = 0.221$ $\theta = 12.8^0$

$\sin\theta = 3\lambda/2d_1 = 0.331$ $\theta = 19.3^0$

Hence, the angles with respect to AA are

$\theta_1 = 6.37 + 45 = 51.4^0$
$\theta_2 = 12.8 + 45 = 57.8^0$
$\theta_3 = 19.3 + 45 = 64.3^0$

---

40. (a) $n\lambda = 2d\sin\theta$ becomes

$d = \lambda/(2\sin\theta) = (0.626\ \text{Å})/(2)(\sin 6.41^0) = 2.80\ \text{Å}$

(b) From Fig. 2.32 there are (4)(1/8) Cl^- and (4)(1/8) Na^+ ions per primitive cell. Thus there is 1/2 of a NaCl molecule per primitive cell. As the mass of a NaCl molecule is (molecular weight of NaCl)/N_A, the density of a primitive cell is

$\rho = (1/2)(\text{molecular weight of NaCl})/d^3N_A$

Solving for N_A yields

$N_A = (1/2)(\text{molecular weight of NaCl})]/d^3\rho$ or

$N_A = (1/2)(58.4\ g)/(2.80 \times 10^{-8}\ cm)^3(2.17\ g/cm^3) = \underline{6.1 \times 10^{23}}$

---

41. (a) $-h\,df = F_G\,dr = [G(hf/c^2)M_S/r^2]\,dr$

$$\int_f^{f'} \frac{df}{f} = -(GM_S/c^2)\int_{R_S}^{\infty} \frac{dr}{r^2}$$

$$\ln(f'/f) = (GM_S/c^2)(1/r)\Big|_{R_S}^{\infty} = -\,GM_S/c^2R_S$$

or $f' = f\exp(-GM_S/c^2R_S)$

(b) For small M_S/R_S, $\exp(-GM_S/c^2R_S) \approx 1 - GM_S/c^2R_S$

so $f' \approx f[1 - GM_S/c^2R_S]$

---

42. (a) The threshold for a black hole is $GM_S/Rc^2 \approx 1$ or

$R \approx GM_S/c^2$

$= (6.67 \times 10^{-11}\ \text{N}\cdot\text{m}^2/\text{kg}^2)(1.99 \times 10^{30}\ \text{kg})/(3.00 \times 10^8\ \text{m/s})^2$

$= 1470\ \text{m}$

This value of ≈ 1500 m should be compared to the normal sun radius of 696,000,000 m

(b) $\dfrac{M_S/4\pi R^3/3}{M_S/4\pi R_S{}^3/3} = R_S{}^3/R^3 = (6.96 \times 10^8)^3/(1.47 \times 10^3)^3 = 1.1 \times 10^{17}$!!

Chapter 3

1. $1/\lambda = R[(1/n^2_f) - (1/n^2_i)]$; For the Balmer series,

   $n_f = 2$; $n_i = 3, 4, 5, \ldots$ The first three lines in the series have wavelengths given by

   $1/\lambda = R(1/2^2 - 1/n^2)$

1st line: $1/\lambda = R(1/4 - 1/9) = (5/36)R$; $\lambda = 36/5R =$ 656.112 nm
2nd line: $1/\lambda = R(1/4 - 1/16) = (3/16)R$; $\lambda = 16/3R =$ 486.009 nm
3rd line: $1/\lambda = R(1/4 - 1/25) = (21/100)R$; $\lambda = 100/21R =$ 433.937 nm

---

2. $1/\lambda = R[1 - (1/n^2)]$ where $n = 2, 3, 4, \ldots$

   and $R = 1.0973732 \times 10^7 \text{ m}^{-1}$;

   For $n = 2$:

   $\lambda = R^{-1}[1 - (1/2^2)]^{-1} = 1.21502 \times 10^{-7}$ m = 121.502 nm (UV)

   For $n = 3$:

   $\lambda = R^{-1}[1 - (1/3^2)]^{-1} = 1.02517 \times 10^{-7}$ m = 102.517 nm (UV)

   For $n = 4$:

   $\lambda = R^{-1}[1 - (1/4^2)]^{-1} = 9.72018 \times 10^{-7}$ m = 972.018 nm (IR)

---

3. a) $\lambda = 102.6$ nm ; $1/\lambda = R[1 - (1/n^2)]$

   $\Rightarrow n = \{R/[R - (1/\lambda)]\}^{1/2}$

   $= \{R/[R - (1/102.6 \times 10^{-9} \text{ m})]\}^{1/2} = 2.99 \approx 3$;

   b) This wavelength cannot belong to either series. Both the Paschen and Brackett series lie in the IR region, while the wavelength of 102.6 nm lies in the UV region.

---

4. $r_n = n^2\hbar^2/mke^2$; where $n = 1, 2, 3, \ldots$

$r_n = n^2(1.055\times10^{-34} \text{ J}\cdot\text{s})^2/[(9.11\times10^{-31} \text{ kg})(9.0\times10^9 \text{ Nm}^2/\text{C}^2)(1.6\times10^{-19} \text{ C})^2]$

$= (0.0529 \text{ nm})n^2$

For $n = 1$: $r_n = 0.0529$ nm

For $n = 2$: $r_n = 0.2121$ nm

For $n = 3$: $r_n = 0.4772$ nm

---

5. (b) The energy levels of a hydrogen-like ion whose charge number is Z is given by

$$E_n = (-13.6\ \text{eV})Z^2/n^2$$

For He+, Z = 2 so we see that the ionization energy (the energy required to take the electron from the state n = 1 to the state n = ∞ is

$$E = (-13.6\ \text{eV})(2^2)/1^2 = -\ \underline{54.4\ \text{eV}}$$

---

6. For Li^{++}, Z = 3 ; From Eq. 3.36

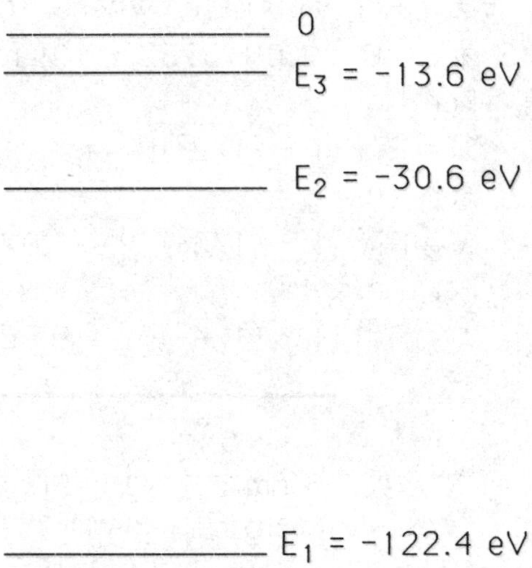

$$E_n = -\ (13.6Z^2)/n^2\ \text{eV}$$

$$= -\ (122.4)/n^2\ \text{eV}$$

So $E_1 = -122.4$ eV

$E_2 = -30.6$ eV

$E_3 = -13.6$ eV

etc.

0

$E_3 = -13.6$ eV

$E_2 = -30.6$ eV

$E_1 = -122.4$ eV

---

7. $r = n^2\hbar^2/Zmke^2 = (n^2/Z)(\hbar^2/mke^2)$;

n = 1

$r = (1/Z)\{(1.055\times10^{-34}\ \text{J·s})^2 \div [(9.11\times10^{-31}\text{kg})(9\times10^9\ \text{N·m}^2/\text{C}^2)(1.6\times10^{-19}\text{C})^2]\}$

$= (5.30\times10^{-11}\ \text{m})/Z$

For He^+, Z = 2

$r = (5.30\times10^{-11}\ \text{m})/2 = 2.65\times10^{-11}\ \text{m} = \underline{0.0265}\ \text{nm}$

For Li^{2+}, Z = 3

$r = (5.30\times10^{-11}\ \text{m})/3 = 1.77\times10^{-11}\ \text{m} = \underline{0.0177}\ \text{nm}$

For Be^{3+}, Z = 4

$r = (5.03\times10^{-11}\ \text{m})/4 = 1.32\times10^{-11}\ \text{m} = \underline{0.0132}\ \text{nm}$

---

8. a) $\Delta E = (13.6 \text{ eV})[(1/1^2) - (1/3^2)] = \underline{12.1} \text{ eV}$

   b) Either $\Delta E = \underline{12.1} \text{ eV}$ or $\Delta E = (13.6 \text{ eV})[(1/1) - (1/2^2)] = \underline{10.2} \text{ eV}$

   and $\Delta E = (13.6 \text{ eV})[(1/2^2) - (1/3^2)] = \underline{1.89} \text{ eV}$

---

9. a) $E_e = (-13.6 \text{ eV})[(1/n_f^2) - (1/n_f^2)]$

   $= (-13.6 \text{ eV})[(1/9) - (1/4)]$

   $= \underline{1.89 \text{ eV}}$

   b) $E = hc/\lambda$

   $\Rightarrow \lambda = hc/E = (4.14 \times 10^{-15} \text{ eV}\cdot\text{s})(3 \times 10^8 \text{ m/s})/(1.89 \text{ eV}) = \underline{658 \text{ nm}}$

   c) $c = \lambda f$

   $\Rightarrow f = c/\lambda = (3 \times 10^8 \text{ m/s})/(657 \times 10^{-9} \text{ m}) = \underline{4.56 \times 10^{14} \text{ Hz}}$

---

10. $E = (-13.6 \text{ eV})[(1/n_f^2) - (1/n_i^2)]$

    a) $E = (-13.6 \text{ eV})[(1/25) - (1/16)] = \underline{0.306 \text{ eV}}$

    b) $E = (-13.6 \text{ eV})[(1/36) - (1/25)] = \underline{0.166 \text{ eV}}$

---

11. a) For the Paschen series; $1/\lambda = R(1/3^2 - 1/n^2)$; the maximum wavelength corresponds to $n_i = 4$

    $1/\lambda_{max} = R(1/3^2 - 1/4^2)$; $\lambda_{max} = \underline{1874.606} \text{ nm}$

    For minimum wavelength, $n_i \rightarrow \infty$

    $1/\lambda_{min} = R(1/3^2 - 1/\infty)$; $\lambda_{min} = 9/R = \underline{820.140} \text{ nm}$

    b) $hc/\lambda_{min} = (hc/1874.606 \text{ nm})/(1.6 \times 10^{-19} \text{ J/eV}) = \underline{0.6627} \text{ eV}$

    $hc/\lambda_{min} = (hc/820.140 \text{ nm})/(1.6 \times 10^{-19} \text{ J/eV}) = \underline{1.515} \text{ eV}$

---

12. $E = K + U = mv^2/2 - ke^2/r$

    But $mv^2/2 = (1/2)ke^2/r$

    Thus $E = (1/2)[-1/(4\pi\varepsilon_0)]\, e^2/r = U/2$, so

    $U = 2E = 2(-13.6 \text{ eV}) = \underline{-27.2} \text{ eV}$ and

    $K = E - U = -13.6 \text{ eV} - (-27.2 \text{ eV}) = \underline{13.6} \text{ eV}$

---

13. a) $r_1 = (0.0529\text{ nm})n^2 = \underline{0.0529}$ nm (when n = 1)

b) $mv = m[ke^2/mr]^{1/2}$

$mv = [(9.1 \times 10^{-3}\text{ kg})(9 \times 10^9\text{ N·m}^2/\text{C}^2)/(5.29 \times 10^{-11}\text{ m})]^{1/2}$
$\times (1.6 \times 10^{-19}\text{ C})$

$mv = \underline{1.99 \times 10^{-24}}$ kg·m/s

c) $L = mvr = (1.99 \times 10^{-24}\text{ kg·m/s})(5.29 \times 10^{-11}\text{ m})$

$L = \underline{1.05 \times 10^{-34}}$ (kg·m²/s)

d) $K = |E| = \underline{13.6}$ eV

e) $U = -2K = \underline{-27.2}$ eV

f) $E = K + U = \underline{-13.6}$ eV

---

14. $\mu \equiv$ reduced mass, $\mu = mM/(m + M) = m/2$ since M is the mass of the positron which equals m the mass of the electron

a) $r_{Pos} = n^2\hbar^2/Z\mu ke^2$

$= n^2\hbar^2/Z(m/2)ke^2 = 2n^2\hbar^2/Zmke^2$

or $r_{Pos} = 2r_{Hyd}$

b) $E_{Pos} = -(\mu k^2e^4/2\hbar^2)(1/n^2)$

$= -[(mk^2e^4/2)/(2\hbar^2)](1/n^2)$

$= -(mk^2e^4/4\hbar^2)(1/n^2)$; n = 1, 2, 3, . . .

or $E_{Pos} = E_{Hyd}/2$

---

15. $hf = \Delta E = [4\pi^2mk^2e^4/(2h^2)]([1/(n-1)]^2 - 1/n^2)$

$f = (2\pi^2mk^2e^4/h^3)((2n-1)/[(n-1)^2\,n^2])$ and as $n \rightarrow \infty$ we have

$f \rightarrow (2\pi^2mk^2e^4/h\varepsilon)(2/n^3)$

The classical frequency is

$f = v/(2\pi r) = [1/(2\pi)](ke^2/m)^{1/2}(1/r^{3/2})$ where $r = n^2h^2/4\pi mke^2$

so $f = [1/(2\pi)](ke^2/m)^{1/2}[8\pi^3e^3mk\sqrt{(mk)}/n^3h^3] = 4\pi^2mk^2e^4/n^3h^3$

---

21. The velocity of fall $v = \Delta y/\Delta t = (0.004\text{ m})/(15.9\text{ s})$
$= \underline{2.52 \times 10^{-4}\text{ m/s}}$

a) The radius, a, is given by:

$$a = \left(\frac{9\zeta v}{2\rho g}\right)^{1/2} = \left(\frac{9}{2}\frac{(1.81 \times 10^{-5}\text{ kg/m·s})(2.52 \times 10^{-4}\text{ m/s})}{(800\text{ kg/m}^3)(9.81\text{ m/s}^2)}\right)^{1/2}$$

$a = \underline{1.62 \times 10^{-6}}$ m

The mass, m, is given by

$m = \rho V = \rho \cdot (4/3)\pi a^3 = (1.33)(800\ kg/m^3)(\pi)(1.62 \times 10^{-6}\ m)^3$
$= \underline{1.42 \times 10^{-14}}$ kg

b) $q = (mg/E)(v + v')/v$ where v is the velocity of fall and v' is the velocity of rise. The electric field is given by

$E = V/d = (4000\ V)/(0.0200\ m) = 200{,}000\ V/m$

$v = 2.52 \times 10^{-4}$ m/s
$v' = (0.004\ m)/(36.0\ s) = 1.11 \times 10^{-4}$ m/s,
$v_1' = 2.31 \times 10^{-4}$ m/s, $v_2' = 1.67 \times 10^{-4}$ m/s,
$v_3' = 3.51 \times 10^{-4}$ m/s, $v_4' = 5.31 \times 10^{-4}$ m/s

$q = (mg/E)(v + v')/v = (6.97 \times 10^{-19}\ C)(2.52 + 1.11)/2.52$
$= 10.0 \times 10^{-19}$ C
$q_1 = (mg/E)(v + v_1')/v = 13.4 \times 10^{-19}$ C

$q_2 = 11.6 \times 10^{-19}$ C
$q_3 = 16.7 \times 10^{-19}$ C
$q_4 = 21.6 \times 10^{-19}$ C

By inspection we choose integers which will yield an elementary charge between 1.5 and 2.0×10^{-19} C

$q/6 = 1.67 \times 10^{-19}$ C
$q_1/8 = 1.68 \times 10^{-19}$ C
$q_2/7 = 1.66 \times 10^{-19}$ C
$q_3/10 = 1.67 \times 10^{-19}$ C
$q_4/13 = 1.66 \times 10^{-19}$ C

The amount of charge gained or lost:

$$q_1 - q = 3.4 \times 10^{-19}\ C$$
$$q_2 - q = 1.6 \times 10^{-19}\ C$$
$$q_1 - q_2 = 1.8 \times 10^{-19}\ C$$
$$q_3 - q = 6.7 \times 10^{-19}\ C$$
$$q_3 - q_2 = 5.1 \times 10^{-19}\ C$$
$$q_3 - q_1 = 3.3 \times 10^{-19}\ C$$
$$q_4 - q = 11.6 \times 10^{-19}\ C$$
$$q_4 - q_3 = 4.9 \times 10^{-19}\ C$$
$$q_4 - q_2 = 10.0 \times 10^{-19}\ C$$

By inspection we again find integers which yield a constant value of e:

$$(q_1 - q)/2 = 1.7$$
$$(q_2 - q)/1 = 1.6$$
$$(q_1 - q_2)/1 = 1.8$$
$$(q_3 - q)/4 = 1.68$$
$$(q_3 - q_2)/3 = 1.70$$
$$(q_3 - q_1)/2 = 1.65$$
$$(q_4 - q)/7 = 1.66$$
$$(q_4 - q_3)/3 = 1.63$$
$$(q_4 - q_2)/6 = 1.67$$

Average of all values = $\underline{1.67 \times 10}^{-19}$ C

---

22. $m = \rho V = (0.9561\ g/cm^3)(4/3)\pi a^3 = 8.418 \times 10^{-11}\ g$

mg/E for use in $q = (mg/E)(v + v')/v$ has the value

$$mg/E = (8.418 \times 10^{-14}\ kg)(9.80\ m/s^2)/(5085\ V/0.01600\ m)$$
$$= 25.96 \times 10^{-19}\ C$$

Using $q = (mg/E)(v + v^1)/v)$ we find the different charges on the drops to be:

$q = (25.96 \times 10^{-19}\ C)(0.85842 + 0.1265)/(0.85842)$
$= 29.78 \times 10^{-19}\ C$

$q_1 = 39.76 \times 10^{-19}\ C$
$q_2 = 28.16 \times 10^{-19}\ C$
$q_3 = 29.84 \times 10^{-19}\ C$
$q_4 = 34.84 \times 10^{-19}\ C$
$q_5 = 36.51 \times 10^{-19}\ C$

$[\lvert\text{Charge differences}\rvert] \times 10^{-19}$ C	$[\lvert\text{Charge differences}\rvert/n] \times 10^{-19}$ C (n chosen by inspection)
$q - q_1 = 9.98$	$9.98/6 = 1.66$
$q - q_2 = 1.62$ (circled)	$1.62/1 = 1.62$
$q - q_3 = 0.06 \approx 0$	------
$q - q_4 = 5.06$	$5.06/3 = 1.69$
$q - q_5 = 6.73$	$6.73/4 = 1.68$
$q_2 - q_1 = 11.6$	$11.6/7 = 1.66$
$q_3 - q_1 = 9.92$	$9.92/6 = 1.65$
$q_4 - q_1 = 4.92$	$4.92/3 = 1.64$
$q_5 - q_1 = 3.25$	$3.25/2 = 1.63$
$q_3 - q_2 = 1.68$ (circled)	$1.68/1 = 1.68$
$q_4 - q_2 = 6.68$	$6.68/4 = 1.67$
$q_5 - q_2 = 8.35$	$8.35/5 = 1.67$
$q_4 - q_3 = 5.00$	$5.00/3 = 1.67$
$q_5 - q_3 = 6.67$	$6.67/4 = 1.66$
$q_5 - q_4 = 1.67$ (circled)	$1.67/1 = 1.67$

Average $q = \underline{1.661 \times 10^{-19}}$ C

24. a) Total charge passed = i·t = (1.00 A)(3600 s) = 3600 C
This is (3600 C)/(1.60×10^{-19} C) = 2.25×10^{22} electrons

As the valence of the copper ion is two, two electrons are required to deposit each ion as a neutral atom on the cathode.

The number of Cu atoms = (number of electrons)/(2)
= $\underline{1.125 \times 10^{22}}$ Cu atoms

b) So the weight (mass) of a Cu atom is:

(1.185 g)/(1.125×10^{22} atoms) = $\underline{1.05 \times 10^{-22}}$ g

c) m = q(molar weight)/(96,500)(2) or
molar weight = m(96,500)(2)/q = (1.185 g)(96,500 C)(2)/(3600 C)
= $\underline{63.53}$ g

---

25. a) $\lambda = C_2 n^2/(n^2 - 2^2)$
so $1/\lambda = (1/C_2)(n^2 - 2^2)/n^2 = (1/C_2)(1 - 2^2/n^2)$
or $1/\lambda = (2^2/C_2)(1/2^2 - 1/n^2) = R(1/2^2 - 1/n^2)$
where $R = (2^2/C_2)$

b) $R = (2^2)/(3645.6 \times 10^{-8}\ \text{cm}) = \underline{109720\ \text{cm}^{-1}}$

---

26. Thomson's device will work for positive or negative particles, so we may apply $q/m \approx V\theta/B^2 \ell d$

a) $q/m \approx V\theta/B^2 \ell d$
= (2000 V)(0.20 radians)/(4.57×10^{-2} T)^2(0.10 m)(0.02 m)
= $\underline{9.58 \times 10^{7}}$ C/kg

b) Since the particle is attracted by the negative plate, it carries a positive charge and is a proton.

c) $v_x = E/B = V/dB$ = (2000 V)/(0.02 m)(4.57×10^{-2} T)
= $\underline{2.19 \times 10^{6}}$ m/s

---

27. $y_1 = (1/2)a_y t^2 = (1/2)(F/m)(\ell/v_x)^2 = (1/2)(Eq/m)(\ell/v_x)^2$
$= (1/2)(V/d)(q/m)(\ell/v_x)^2$

$y_2 = D\tan\theta = Dv_y/v_x = Da_y t/v_x = D(Eq/m)(\ell/v_x)(1/v_x)$

$= D(V/d)(q/m)(\ell/v_x^2)$

$Y = y_1 + y_2 = (V/d)(q/m)(\ell/v_x^2)(\ell/2 + D)$

Solving for q/m yields

$(d/V)(Y)(v_x^2/\ell)/[(\lambda/2) + D] = q/m$

or $q/m = (Yv_x^2 d)/[V\ell(\ell/2) + D]$

---

28. a) From equation 3.15 we have

$\Delta n \propto (\sin\phi/2)^{-4}$ or $\Delta n_2 = \Delta n_1 (\sin\phi_1/2)^4/(\sin\phi_2/2)^4$

Thus the number of α's scattered at 40 degrees is given by

$\Delta n_2 = (100\text{ cpm})[\sin(20/2)]^4/[\sin(40/2)]^4$
$= (100\text{ cpm})(\sin 10/\sin 20)^4 = \underline{6.64}\text{ cpm}$

Similarly

Δn at 60 degrees = $\underline{1.45}$ cpm
Δn at 80 degrees = $\underline{0.533}$ cpm
Δn at 100 degrees = $\underline{0.264}$ cpm

b) From 3.15 doubling $(1/2)m_\alpha v_\alpha^2$ reduces Δn by a factor of 4.
Thus Δn at 20 degrees = (1/4)(100 cpm) = $\underline{25}$ cpm

c) From 3.15 we find

$(\Delta n_{Cu})/(\Delta n_{Au}) = (Z^2_{Cu} N_{Cu})/(Z^2_{Au} N_{Au})$

$Z_{Cu} = 29 \qquad Z_{Au} = 79$

N_{Cu} = number of Cu nuclei per unit area
= number of Cu nuclei per unit volume · foil thickness
$= [(8.9\ g/cm^3)\cdot(6.02 \times 10^{23}\ nuclei)/(63.54\ g)]\cdot t$
$= 8.43 \times 10^{22} t$

$N_{Au} = [(19.3\ g/cm^3)\cdot(6.02 \times 10^{23}\ nuclei)/(197.0\ g)]\cdot t$
$= 5.90 \times 10^{22} t$

So $\Delta n_{Cu} = \Delta n_{Au}\,(29)^2(8.43 \times 10^{22})/(79)^2(5.90 \times 10^{22})$
$= (100)(29/79)^2(8.43/5.90)$
$= \underline{19.3\ cpm}$

---

29. The initial energy of the system of α plus copper nucleus is 13.9 MeV and is just the kinetic energy of the α when the α is far from the nucleus. The final energy of the system may be evaluated at the point of closest approach when the kinetic energy is zero and the potential energy is $(2e)(Ze)/(4\pi\varepsilon_0 r)$ where r is approximately equal to the nuclear radius of copper. Invoking conservation of energy

$$E_i = E_f$$

$$K_\alpha = (2Ze^2)/(4\pi\varepsilon_0 r)$$

or

$$r = (2Ze^2)/(4\pi\varepsilon_0 K_\alpha)$$
$$= [(2)(29)(1.60 \times 10^{-19})^2(8.99 \times 10^9)] \div [(13.9 \times 10^6\ eV)(1.60 \times 10^{-19}\ J/eV)]$$
$$= \underline{6.00 \times 10^{-15}}\ m$$

Chapter 4

1. $\lambda = h/p = h/mv = (6.63 \times 10^{-34})/(1.67 \times 10^{-27})(10^{6}) = \underline{3.97 \times 10}^{-13}$ m

2. $\lambda = h/p = 6.63 \times 10^{-34}/(1.67 \times 10^{-27} \times 10^{6}) = \underline{3.97 \times 10}^{-13}$ m

3. $\lambda = h/p = h/mv = (6.63 \times 10^{-34})/(75)(5) = \underline{1.77 \times 10}^{-36}$ m

4. Taking $\lambda = 0.1$ nm, and using $p = h/\lambda = mv$, we get

   $v = h/m\lambda = (6.626 \times 10^{-34}\ \text{J·s})/(9.11 \times 10^{-31}\ \text{kg})(0.1 \times 10^{-9}\ \text{m})$

   $= \underline{7.28 \times 10}^{6}$ m

6. $p = (2mK)^{1/2}$ and $\lambda = h/p = h/(2mK)^{1/2}$

   $\lambda = (6.626 \times 10^{-34})/(2 \times 1.67 \times 10^{-27} \times 10 \times 10^{6} \times 1.6 \times 10^{-19})^{1/2}$

   $\lambda = \underline{9.06 \times 10}^{-15}$ m

7. $\lambda = h/p = h/(2mK)^{1/2} = h/(2meV)^{1/2} = [h/(2me)^{1/2}]\ V^{-1/2}$

   $\lambda = [(6.63 \times 10^{-34})/(2 \times 9.1 \times 10^{-31} \times 1.6 \times 10^{-19})^{1/2}]\ V^{-1/2}$

   $\lambda = \underline{1.226}\ V^{-1/2}$ nm

8. $m = 0.2$ kg; $mgh = mv^{2}/2$; $v = (2gh)^{1/2}$

   $p = mv = m(2gh)^{1/2} = (0.2)[2(9.80)(50)]^{1/2} = 6.261$ kg·m/s

   $\lambda = h/p = (6.63 \times 10^{-34})/6.261 = \underline{1.06 \times 10}^{-34}$ m

Chapter 4

9. Since $\lambda = 2a_0 = 2(0.0529)$ nm $= 0.1008$ nm, the energy of the electron is nonrelativistic, so we can use $p = h/\lambda$ and $K = p^2/2m$;

$K = h^2/2m\lambda^2 = (6.626 \times 10^{-34})^2/[2(9.11 \times 10^{-31})(1.008 \times 10^{-10})^2]$

$K = 21.5 \times 10^{-18}$ J $= \underline{134}$ eV

This is about ten times as large as the ground-state energy of hydrogen, which is 13.6 eV.

---

10. (a) In this problem, the electron must be treated relativistically.

The momentum of the electron is

$p = h/\lambda = (6.626 \times 10^{-34}$ J·s$)/10^{-14}$ m $= 6.626 \times 10^{-20}$ kg·m/s

The energy of the electron is

$E = (p^2c^2 + m^2c^4)^{1/2}$

$E = [(6.626 \times 10^{-20})^2(3 \times 10)^2 + (0.511 \times 10^6)^2(1.6 \times 10^{-19})^2]^{1/2}$

$E = 1.99 \times 10^{-11}$ J $= 1.24 \times 10^8$ eV

so that $K = E - mc^2 \approx \underline{124}$ MeV

(b) The kinetic energy is too large to expect that the electron could be confined to a region the size of the nucleus.

---

11. Using $p = h/\lambda = mv$, we find that

$$v = h/m\lambda = (6.626 \times 10^{-34})/(9.11 \times 10^{-31})(1 \times 10^{-10}) = 7.27 \times 10^{6} \text{ m/s}$$

From the principle of conservation of energy, we get

$$eV = mv^2/2 = (9.11 \times 10^{-31})(7.27 \times 10^{6})^2/2 = 2.41 \times 10^{-17}$$

$$= 151 \text{ eV} ; \quad \text{Therefore, } V = \underline{151 \text{ V}}$$

---

12. $\sin\theta = n\lambda/a$ and $\sin\theta \approx \tan\theta = x/L$

$x_1/L = \lambda/a$ and $x_2/L = 2\lambda/a \Rightarrow x_2 - x_1/L = \lambda/a$ or

$\lambda = a\Delta x/L = (5 \text{ Å} \times 2.1 \text{ cm})/(20 \text{ cm}) = 0.525 \text{ Å}$

$E = p^2/2m = L^2/2m\lambda^2 = (hc)^2/2mc^2\lambda^2$

$= (1.24 \times 10^4 \text{ eV·Å})^2/2(5.11 \times 10^5 \text{ eV})(0.525 \text{ Å})^2 = \underline{546} \text{ eV}$

---

13. (a) $\lambda = h/mv = (6.63 \times 10^{-34})/(1.67 \times 10^{-27})(0.4) = \underline{9.93 \times 10^{-7}} \text{ m}$

(b) Substituting $\phi = \pi$ into Eq. 4.29 gives

$$\theta = \arcsin(\lambda/2d) = 0.0284°$$

$y/D = \tan\theta$ so $y = D\tan\theta = (10 \text{ m})(\tan 0.0284°) = \underline{4.97} \text{ mm}$

(c) We cannot say the neutron passed through one slit. We can only say it passed through the slits.

---

14. $\Delta x \Delta p = \hbar$ where

$$\Delta p = m\Delta v = (0.05 \text{ kg})(10^{-3} \times 30 \text{ m/s}) = 1.5 \times 10^{-3} \text{ kg·m/s}$$

Therefore,

$$\Delta x = \hbar/2\Delta p = (6.626 \times 10^{-34} \text{ J·s})/4\pi(1.5 \times 10^{-3} \text{ kg·m/s})$$

$$\Delta x = \underline{3.51 \times 10^{-32}} \text{ m}$$

---

15. $K = mv^2/2 = p^2/2m$;

$$(1 \times 10^6)(1.6 \times 10^{-19}) = p^2/2(1.67 \times 10^{-27}) \Rightarrow p = 2.312 \times 10^{-20} \text{ kg·m/s}$$

$\Delta p = 0.05p = 1.160 \times 10^{-21}$ kg·m/s and $\Delta x \cdot \Delta p = h/4\pi$, thus

$$\Delta x = (6.63 \times 10^{-34})/(1.16 \times 10^{-21})(4\pi) = \underline{4.56 \times 10^{-14}} \text{ m}$$

Note that non-relativistic treatment has been used, which is justified since the KE is only

$$(1.6 \times 10^{-13}) \times 100\%/(1.50 \times 10^{-10}) = 0.11\%$$

of the rest mass energy.

---

16. $\Delta x \cdot \Delta p = h/4\pi \Rightarrow \Delta p = h/4\pi\Delta x \Rightarrow \Delta v = h/4\pi m\Delta x$

$$\Delta v = (6.63 \times 10^{-34})/(4\pi \times 9.1 \times 10^{-31} \times 10^{-10}) = \underline{5.79 \times 10^5} \text{ m/s}$$

---

17. $\Delta y/x = \Delta p_y/p_x$ and $d\Delta p_y = h$. Eliminate Δp_y and solve for x.

$$x = p_x(\Delta y)d/h$$

$$= (10^{-3}\ \text{kg})(100\ \text{m/s})(10^{-2}\ \text{m})(2 \times 10^{-3}\ \text{m})/(6.63 \times 10^{-34}\ \text{J·s})$$

$$= \underline{3 \times 10^{27}}\ \text{m}$$

This is 15 times greater than the diameter of the universe.

---

18. With <u>one</u> slit open $P_1 = |\Psi_1|^2$ or $P_2 = |\Psi_2|^2$

With <u>both</u> slits open, $P = |\Psi_1 + \Psi_2|^2$

At a maximum, the wavefunctions are in phase so

$$P_{max} = (|\Psi_1| + |\Psi_2|)^2$$

At a minimum, the wavefunctions are out of phase and

$$P_{min} = (|\Psi_1| - |\Psi_2|)^2$$

Now $P_1/P_2 = |\Psi_1|^2/|\Psi_2|^2 = 25$ so $|\Psi_1|/|\Psi_2| = 5$, and

$$P_{max}/P_{min} = (|\Psi_1| + |\Psi_2|)^2/(|\Psi_1| - |\Psi_2|)^2$$

$$= (5|\Psi_2| + |\Psi_2|)^2/(5|\Psi_2| - |\Psi_2|)^2 = 6^2/4^2 = 36/16$$

$$= \underline{2.25}$$

---

19. The relativistic expression for the kinetic is given by

$$K = (\gamma - 1)mc^2$$

where $mc^2 = 0.511$ MeV for the electron. Solving this for γ gives

$$\gamma = (K/mc^2) + 1$$

Chapter 4

Using the three values of K given, we find that

(a) For $K = 0.01$ MeV, $\gamma_1 = 1.02$

(b) For $K = 1$ MeV, $\gamma_2 = 2.96$

(c) For $K = 100$ MeV, $\gamma_3 = 197$

Since $\gamma = 1/(1 - v^2/c^2)^{1/2}$, we can use the above results to calculate the corresponding values of v. Solving for v gives

$$v = c(1 - 1/\gamma^2)^{1/2}$$

(a) For $\gamma_1 = 1.02$, $v_1 = 5.91 \times 10^7$ m/s

(b) For $\gamma_2 = 2.96$, $v_2 = 2.82 \times 10^8$ m/s

(c) For $\gamma_3 = 197$, $v_3 = 3.00 \times 10^8$ m/s

The corresponding values of the momentum, using $p = \gamma mv$, are

$$p_1 = 5.49 \times 10^{-23} \text{ kg·m/s}$$

$$p_2 = 7.60 \times 10^{-22} \text{ kg·m/s}$$

$$p_3 = 5.38 \times 10^{-20} \text{ kg·m/s}$$

We can use the uncertainty relation, $\Delta x \Delta p > \hbar/2$, with $\Delta x = a$, to determine the width of the slit. Since it is required to resolve a 1% difference in momentum, we take $\Delta p = 0.01p$. This gives

$$\Delta x \Delta p = a(0.01p) = \hbar/2 \quad \text{or} \quad a = \hbar/2(0.01p)$$

Using the calculated values of p, we find

(a) For $p = p_1$, $a_1 = \underline{0.0961}$ nm

(b) For $p = p_2$, $a_2 = \underline{0.00694}$ nm

(c) For $p = p_3$, $a_3 = \underline{9.80 \times 10^{-14}}$ m

---

20. (a) $E_\gamma = hc/\lambda = (6.626 \times 10^{-34}\ \text{J·s})(3 \times 10^8\ \text{m/s})$

$$\div(6.944 \times 10^{-7}\ \text{m})(1.6 \times 10^{-19}\ \text{J/eV}) = \underline{1.789}\ \text{eV}$$

(b) $E_{Total} = (1.789\ \text{eV})(1.6 \times 10^{-19}\ \text{J/eV})(3 \times 10^{19}) = \underline{8.59}\ \text{J}$

(c) Not all ions participate in the process of stimulated emission.

---

21. (a) $\Delta x \Delta p = \hbar$ so if $\Delta x = r$, $\Delta p \approx \hbar/r$

(b) $K = p^2/2m \approx (\Delta p)^2/2m = \hbar^2/2mr^2$

$U = -\,ke^2/r$

$E = \hbar^2/2mr^2 - ke^2/r$

(c) To minimize E, take $dE/dr = -\,\hbar^2/mr^3 + ke^2/r^2 = 0$

$$\rightarrow\ r = \hbar^2/mke^2 = \text{Bohr radius} = a_0$$

Then $E = (\hbar^2/2m)(mke^2/\hbar^2) - ke^2(mke^2/\hbar^2) = -\,mk^2e^4/2\hbar^2$

$E = \underline{-13.6}$ eV

---

23. (a) $n\lambda = d \sin\phi$ or

$$\sin\phi = \frac{n\lambda}{d} = \frac{n}{d}\frac{h}{p} = \frac{n}{d}\frac{h}{(2m_0K)^{1/2}} = \frac{nhc}{d(2m_0c^2K)}$$

(b) $d_1 = \dfrac{nhc}{(\sin\phi)(2m_0c^2K)^{1/2}}$

$= \dfrac{(1)(12.41 \times 10^{-7} \text{ eV·m})}{(\sin 24.1°)(2 \times 0.511 \times 10^6 \text{ eV} \times 100 \text{ eV})^{1/2}}$

$= 3.00 \times 10^{-10}$ m = $\underline{3.00 \text{ Å}}$

$d_2 = \dfrac{(2)(12.41 \times 10^{-7} \text{ eV·m})}{(\sin 54.9°)(2 \times 0.511 \times 10^6 \text{ eV} \times 100 \text{ eV})^{1/2}}$

$= 3.00 \times 10^{-10}$ m

Since we obtain the same spacing in both cases, 24.1° must correspond to n = 1 and 54.9° to n = 2.

---

25. (a) $g(\omega) = (2\pi)^{-1/2}\int_{-\infty}^{+\infty} V(t)(\cos\omega t - i\sin\omega t)\,dt$

$V(t)\sin\omega t$ is an odd function so this integral vanishes leaving

$g(\omega) = 2(2\pi)^{-1/2}\int_0^{\tau} V_0 \cos\omega t\,dt = (2/\pi)^{1/2} V_0(\sin\omega\tau)/\omega$

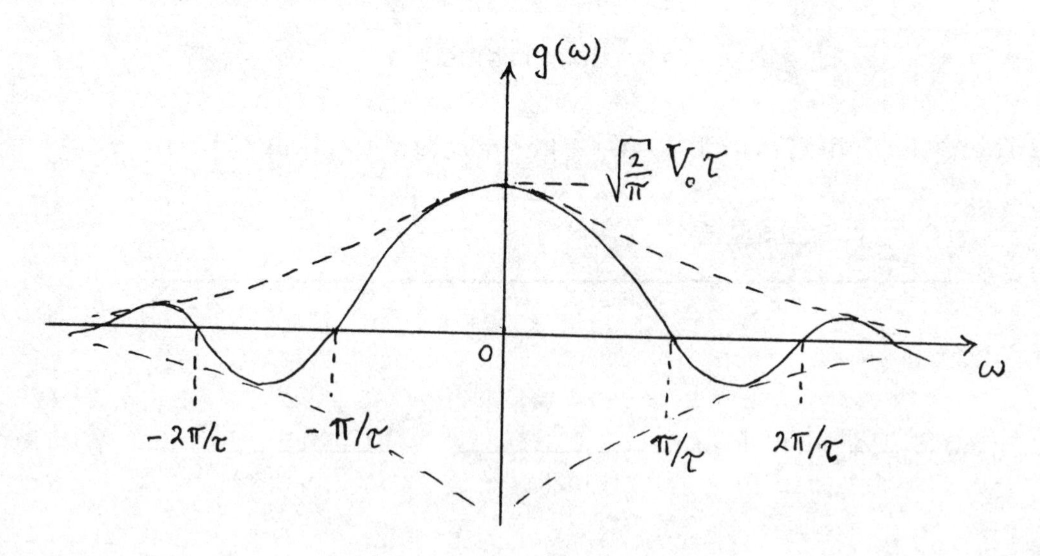

(b) Since the major contribution to this pulse comes from ω's between $-\pi/\tau$ and π/τ, let $\Delta\omega \approx 2\pi/\tau$, and since $\Delta t = 2\tau$,

$$\Delta\omega\Delta t \approx (2\pi/\tau)\cdot 2\tau = \underline{4\pi}$$

(c) Substituting $\Delta t = 1\ \mu s$ in $\Delta\omega = 4\pi/\Delta t$ we find

$$\Delta f = 2/\Delta t = 2/(10^{-6}\ s) = \underline{2 \times 10^6}\ Hz$$

For $\Delta t = 1$ ns,

$$\Delta f = 2/(10^{-9}\ s) = \underline{2 \times 10^9}\ Hz$$

---

26. For a **free, non-relativistic** electron

$$E = mv^2/2 = p^2/2m$$

Since the wavelength and frequency of the electron's deBroglie wave are given by $p = \hbar k$ and $E = \hbar\omega$, substituting these results gives the *dispersion relation*

$$\omega = \hbar k^2/2m$$

So

$$v_g = d\omega/dk = \hbar k/m = p/m = v_0$$

---

27. (a) $f(x) = (2\pi)^{-1/2}\int_{-\infty}^{+\infty} g(k)e^{ikx}\,dk = A'(2\pi)^{-1/2}\int_{-\infty}^{+\infty} e^{-(k-k_0)^2/2(\Delta k)^2}e^{ikx}\,dk$

$$= A'(2\pi)^{-1/2}\int_{-\infty}^{+\infty} e^{(-k^2 + 2k_0k - k_0^2)/2(\Delta k)^2 + ikx}\,dk$$

$$f(x) = A'(2\pi)^{-1/2}\, e^{-k_0^2/2(\Delta k)^2} \int_{-\infty}^{+\infty} e^{[-k^2 + (2k_0 + 2ix(\Delta k)^2)k]/2(\Delta k)^2}\, dk$$

In order to put this integral into the standard form of $\int_{-\infty}^{+\infty} e^{-az^2} dz$,

we complete the square, a humble but quite useful technique. Thus

$$f(x) = A'(2\pi)^{-1/2}\, e^{-k_0^2/2(\Delta k)^2 + \{k_0 + ix(\Delta k)^2\}^2/2(\Delta k)^2}$$

$$\times \int_{-\infty}^{+\infty} e^{[-k^2 + \{2k_0 + 2ix(\Delta k)^2\}k - \{k_0 + ix(\Delta k)\}^2]/2(\Delta k)^2}\, dk$$

$$= A'(2\pi)^{-1/2}\, e^{ik_0x - x^2(\Delta k)^2/2} \int_{-\infty}^{+\infty} e^{-[k - \{k_0 + ix(\Delta k)^2\}]^2/2(\Delta k)^2}\, dk$$

$$= A'(2\pi)^{-1/2}\, e^{-x^2(\Delta k)^2/2}\, e^{ik_0x} \int_{-\infty}^{+\infty} e^{-z^2/2(\Delta k)^2}\, dz$$

where $z = k - \{k_0 + ix(\Delta k)^2\}$.

As $\int_{-\infty}^{+\infty} e^{-z^2/2(\Delta k)^2}\, dz = (2\pi)^{1/2}\Delta k$, $\quad f(x) = A e^{-x^2(\Delta k)^2/2}\, e^{ik_0x}$ . . .

This is just a gaussian envelope multiplying a harmonic wave with wave number k_0. A plot of the real part of $f(x)$ is shown on the following page.

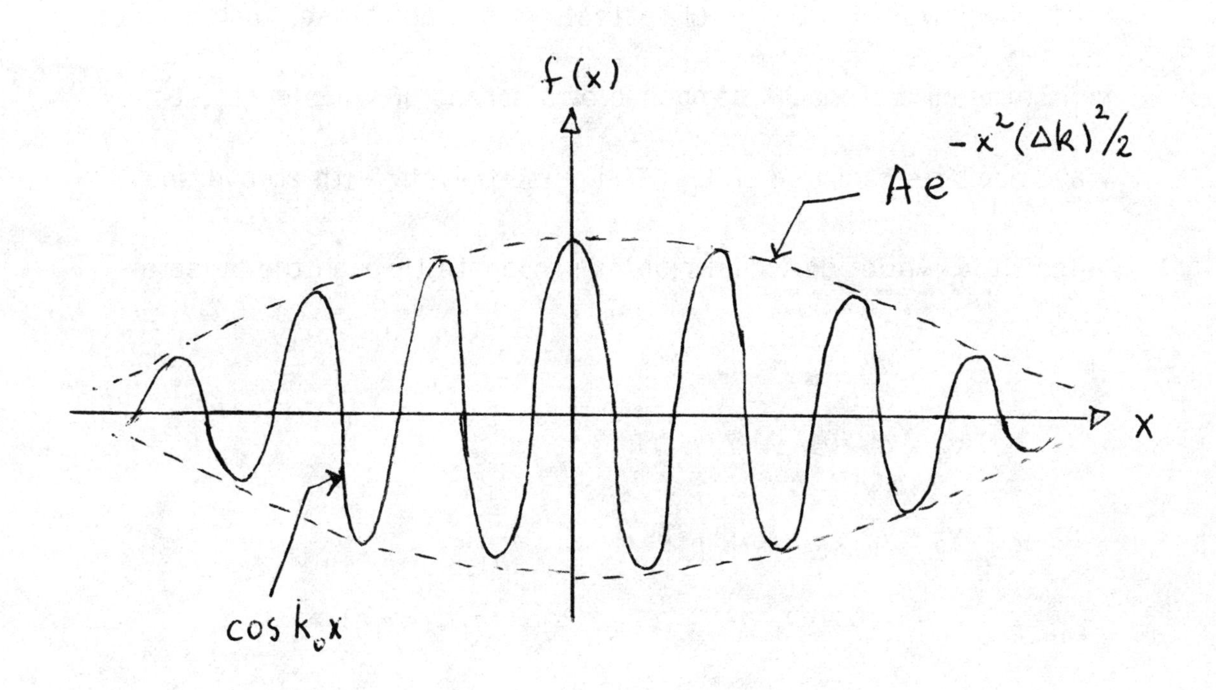

(b) Comparing $Ae^{-x^2(\Delta k)^2/2}$ to $Ae^{-x^2/2(\Delta x)^2}$

where Δx is the width of f(x),

$$(\Delta k)^2/2 = 1/2(\Delta x)^2 \quad \text{or} \quad \Delta x \Delta k = 1$$

(c) As $p = \hbar k$, $\quad \Delta p = \hbar \Delta k \quad$ or $\quad \Delta k = \Delta p/\hbar$

so $\quad \Delta x \Delta k = 1 \quad$ becomes $\quad \Delta x \Delta p = \hbar$

---

28. $v_p = (2\pi S/\lambda\rho)^{1/2}$ means that individual harmonic waves (ripples) with short wavelength travel fastest. A surface pulse, such as that generated by a stone, is composed of a spread of wavelengths or wave numbers centered on k_0. The pulse travels with an average velocity v_g while individual ripples propagate through the pulse at v_p. Writing

$$v_p = k^{1/2}(S/\rho)^{1/2}$$

$$v_g = v_p\big|_{k_0} + k\,(dv_p/dk)\big|_{k_0}$$

$$= k_o^{1/2}(S/\rho)^{1/2} + (k_o)(1/2)(k_o)^{-1/2}(S/\rho)^{1/2}$$

$$= (3/2)k_o^{1/2}(S/\rho)^{1/2}$$

Assuming that the dominant individual wave has $k = k_0$, $v_p = k_0^{1/2}(S/\rho)^{1/2}$. Hence $v_g > v_p$ and the individual ripples move inward through the disturbance.

Chapter 5

1. a) $A \sin(2\pi x/\lambda) = A \sin(5 \times 10^{10} x)$

   so $2\pi/\lambda = 5 \times 10^{10}\ m^{-1}$, $\lambda = 2\pi/(5 \times 10^{10}) = \underline{1.26 \times 10^{-10}}$ m

   c) $K = p^2/2m$ $\quad m = 9.11 \times 10^{-31}$ kg

   $K = (5.26 \times 10^{-24}\ kg{\cdot}m/s)^2/(2 \times 9.11 \times 10^{-31}\ kg) = 1.52 \times 10^{-17}$ J

   $K = (1.52 \times 10^{-17}\ J)/(1.6 \times 10^{-19}\ J/eV) = \underline{95}$ eV

---

2. $\psi(x) = A \cos kx + B \sin kx$

   $\partial\psi/\partial x = - kA \sin kx + kB \cos kx$

   $\partial^2\psi/\partial x^2 = - k^2A \cos kx - k^2B \sin kx$

   $(- 2m/\hbar^2)(E - U)\psi = (- 2mE/\hbar^2)(A \cos kx + B \sin kx)$

   Therefore the Schrodinger equation is satisfied if

   $\partial^2\psi/\partial x^2 = (- 2m/\hbar^2)(E - U)\psi$ or

   $- k^2(A \cos kx + B \sin kx) = (- 2mE/\hbar^2)(A \cos kx + B \sin kx)$

   Therefore $E = \underline{\hbar^2k^2/2m}$

---

3. Solving the Schrodinger equation for U with E = 0 gives

   $U = (\hbar^2/2m)(d^2\psi/dx^2)/\psi$

   If $\psi = Ae^{-x^2/L^2}$, then $d^2\psi/dx^2 = (4Ax^3 - 6AxL^2)e^{-x^2/L^2}/L^4$

   $U = (\hbar^2/2mL^2)(4x^2/L^2 - 6)$

---

4. $E = n^2h^2/8mL^2 \rightarrow n^2 = 8mL^2E/h^2$

   But $E = mv^2/2$ so $n^2 = 4m^2L^2v^2/h^2$ or $n = 2mLv/h$

   Now $v = 1$ Å/year $= 3.17 \times 10^{-8}$ Å/s so

   $n = 2(0.005\ kg)(0.2\ m)(3.17 \times 10^{-18}\ m/s)/(6.63 \times 10^{-34}\ J{\cdot}s) = \underline{9.6{\times}10^{12}}$

Chapter 5

5. $E_n = (h^2/8mL^2)n^2 = 0$ As E is totally K.E.,

$E = 0 = p_x^2/2m$ or $p_x \equiv 0$ which means $\Delta p_x = 0$

Consequently $\Delta p_x \Delta x = h \rightarrow \Delta x \rightarrow \infty$ which is impossible since

Δx can, at most, equal L.

---

6. $E_n = n^2h^2/8mL^2$

$\Delta E = E_2 - E_1 = 3h^2/8mL^2 = 3(hc)^2/8mc^2L^2$ and $\Delta E = hf = hc/\lambda$

Hence $\lambda = 8mc^2L^2/3hc = 8(938 \times 10^6 \text{ eV})(10^{-4} \text{ Å})^2/[3(12400 \text{ eV·Å})]$

$\lambda = \underline{2.02 \times 10^{-3}} \text{ Å} = \underline{2.02 \times 10^{-4}} \text{ nm}$ (γ-ray)

7.

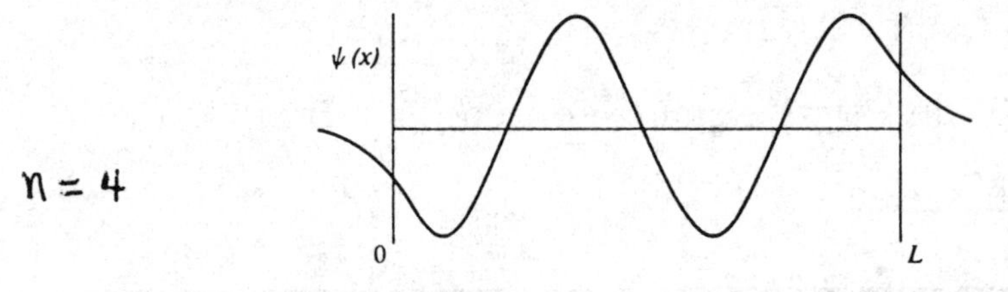

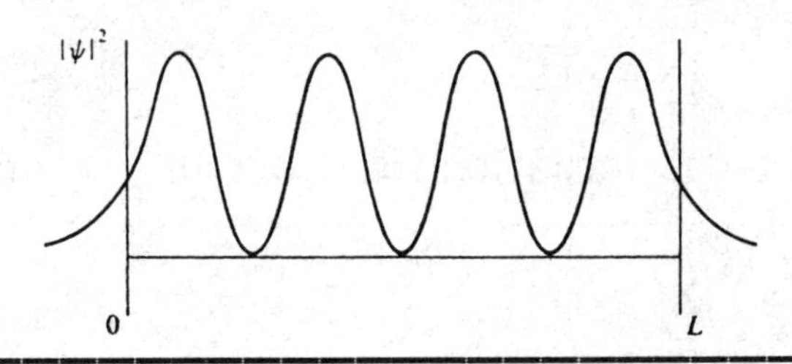

---

8. $E_n = n^2h^2/8mL^2$

$h^2/8mL^2 = (6.63 \times 10^{-34})^2/8(9.11 \times 10^{-31})(10^{-10})^2 = 6.03 \times 10^{-18}$ J

$= 37.7$ eV

(a) $E_1 = \underline{37.7}$ eV

$E_2 = 37.7 \times 2^2 = \underline{151}$ eV

$E_3 = 37.7 \times 3^2 = \underline{339}$ eV

$E_4 = 37.7 \times 4^2 = \underline{603}$ eV

Chapter 5

(b) $hf = hc/\lambda = E_{n_i} - E_{n_f}$

$\lambda = hc/(E_{n_i} - E_{n_f}) = (1243 \text{ eV·nm})/(E_{n_i} - E_{n_f})$

For $n_i = 4$, $n_f = 1$,

$E_{n_i} - E_{n_f} = 603 \text{ eV} - 37.7 \text{ eV} = 565 \text{ eV}$, $\lambda = \underline{2.2}$ nm

$n_i = 4$, $n_f = 2$, $\lambda = \underline{2.75}$ nm

$n_i = 4$, $n_f = 3$, $\lambda = \underline{4.71}$ nm

$n_i = 3$, $n_f = 1$, $\lambda = \underline{4.12}$ nm

$n_i = 3$, $n_f = 2$, $\lambda = \underline{6.59}$ nm

$n_i = 2$, $n_f = 1$, $\lambda = \underline{11}$ nm

---

9. $\psi_1(x) = (2/L)^{1/2} \cos(\pi x/L)$; $P_1(x) = (2/L) \cos^2(\pi x/L)$

$\psi_2(x) = (2/L)^{1/2} \sin(2\pi x/L)$; $P_2(x) = (2/L) \sin^2(2\pi x/L)$

$\psi_3(x) = (2/L)^{1/2} \cos(3\pi x/L)$; $P_3(x) = (2/L) \cos^2(3\pi x/L)$

---

10. $\Delta E = hc/\lambda = (h^2/8mL^2)[2^2 - 1^2]$

and $L = [(3/8)\, h\lambda/mc]^{1/2} = 7.93 \times 10^{-10} = \underline{7.93\ \text{Å}}$

---

11. (a) Proton in a box of width $L = 2$ **Å**

$E_1 = h^2/8m_pL^2 = (6.626 \times 10^{-34} \text{ J·s})^2/8(1.67 \times 10^{-27}\text{kg})(2 \times 10^{-10} \text{ m})^2$

$= \underline{8.22 \times 10}^{-22} \text{ J} = (8.22 \times 10^{-22} \text{ J})/(1.6 \times 10^{-19} \text{ J/eV}$

$= \underline{5.13 \times 10}^{-3} \text{ eV}$

(b) Electron in the same box :

$E_1 = h^2/8m_eL^2 = (6.626 \times 10^{-34} \text{ J·s})^2/8(9.11 \times 10^{-31}\text{kg})(2 \times 10^{-10} \text{ m})^2$

$= 1.506 \times 10^{-18} \text{ J} = \underline{9.40 \text{ eV}}$

(c) The electron has a much higher energy because it is much less massive .

---

12. (a) Still, $n\lambda/2 = L$ so $p = h/\lambda = nh/2L$

$K = [c^2p^2 + (m_0c^2)^2]^{1/2} - (m_0c^2) = E - m_0c^2$

$E_n = [(nhc/2L)^2 + (m_0c^2)^2]^{1/2}$,

$K_n = [(nhc/2L)^2 + (m_0c^2)^2]^{1/2} - m_0c^2$

(b) Taking $L = 10^{-12}$ m, $m_0 = 9.11 \times 10^{-31}$ kg, and $n = 1$ we find

$K_1 = \underline{4.69 \times 10^{-14}}$ J

Nonrel. $E_1 = h^2/8mL^2 = (6.63 \times 10^{-34})^2/(8)(9.11 \times 10^{-31})(10^{-24})$

$= \underline{6.03 \times 10^{-14}}$ J

Comparing this with K_1, we see that this value is too big by 29%.

---

13. (a) $U = (e^2/4\pi\varepsilon_0 d)\,[\,-1 + 1/2 - 1/3 + (-1 + 1/2) + (-1)]$

$= (-7/3)e^2/(4\pi\varepsilon_0 d) = (-7/3)ke^2/d$

(b) $K = 2E_1 = 2h^2/(8m \times 0.9d^2) = h^2/36md^2$

(c) $E = U + K$ and $dE/dd = 0$ for a minimum

$[(+7/3)e^2k/d^2] - h^2/18md^3 = 0$

$d = 3h^2/(7)(18ke^2m)$ or $d = h^2/42mke^2$

$d = (6.63 \times 10^{-34})^2/(42)(9.11 \times 10^{-31})(9 \times 10^9)(1.6 \times 10^{-19}\text{ C})^2$

$= 0.5 \times 10^{-10}$ m $= \underline{0.050 \text{ nm}}$

(d) Since the lithium spacing is a, where $Na^3 = V$, and the density is Nm/V, where m is the mass of one atom, we get

$a = (Vm/Nm)^{1/3} = (m/\text{density})^{1/3}$

$a = (1.66 \times 10^{-27}\text{ kg} \times 7/530\text{ kg})^{1/3}$ m $= 2.8 \times 10^{-10}$ m $= \underline{0.28 \text{ nm}}$

(2.8 times larger than 2d)

Chapter 5

14. (a) $\psi(x) = A \sin(\pi x/L)$, $L = 3$ **Å**

$$\text{Normalization} = 1 = \int_0^L |\psi|^2\, dx = \int_0^L A^2 \sin^2(\pi x/L)\, dx = LA^2/2$$

so $A = (2/L)^{1/2}$

$$P(1\ \text{Å}) = \int_0^{L/3} |\psi|^2\, dx = (2/L) \int_0^{L/3} \sin^2(\pi x/L)\, dx$$

$$P(1\ \text{Å}) = (2/\pi) \int_0^{\pi/3} \sin^2 \phi\, d\phi = (2/\pi)[(\pi/6) - (3)^{1/2}/8] = \underline{0.1955}$$

(b) $\psi = A \sin(100\pi x/L)$, $A = (2/L)^{1/2}$

$$P(1\ \text{Å}) = (2/L) \int_0^{L/3} \sin^2(100\pi x/L)\, dx$$

$$= (2/L)(L/100\pi) \int_0^{100\pi/3} \sin^2 \phi\, d\phi$$

$$= (1/50\pi)[(100\pi/6) - \cos(200\pi/3)/4]$$

$$= (1/3) - \cos(2\pi/3)/200\pi$$

$$= (1/3) - \sqrt{3}/400\pi = \underline{0.3326}$$

(c) Yes: At large quantum numbers the probability approaches 1/3.

Chapter 5

15. (a) The wavefunctions and probability densities are the same as those shown in the two lower curves in Figure 5.9 of the text.

(b) $$P_1 = \int_{1.5\ \text{Å}}^{3.5\ \text{Å}} |\psi|^2\, dx = (2/10\ \text{Å}) \int_{1.5\ \text{Å}}^{3.5\ \text{Å}} \sin^2(\pi x/10)\, dx$$

$$= (1/5)\Big[(x/2) - (10/4\pi)\sin(\pi x/5)\Big]_{1.5}^{3.5}$$

In the above result we used $\int \sin^2 ax\, dx = (x/2) - (1/4a)\sin(2ax)$

Therefore, $$P_1 = (1/10)\Big[x - (5/\pi)\sin(\pi x/5)\Big]_{1.5}^{3.5}$$

$$= (1/10)\{3.5 - (5/\pi)\sin[(\pi)(3.5)/5] - 1.5 + (5/\pi)\sin[\pi(1.5)/5]\}$$

$$= (1/10)[2.0 + (5/\pi)(\sin 0.3\pi - \sin 0.7\pi)] = (1/10)[2.00 - 0.0]$$

$P_1 = \underline{0.200}$

(c) $$P_2 = (1/5) \int_{1.5}^{3.5} \sin^2(\pi x/5)\, dx = (1/5)\Big[x/2 - (5/4\pi)\sin(0.4\pi x)\Big]_{1.5}^{3.5}$$

$$P_2 = (1/10)\Big[x - (5/2\pi)\sin(0.4\pi x)\Big]_{1.5}^{3.5}$$

$$P_2 = (1/10)[2.0 + (0.798)(\sin 0.4\pi \times 1.5 - \sin 0.4\pi \times 3.5)] = \underline{0.351}$$

(d) Using $E = n^2h^2/8mL^2$ we find $E_1 = \underline{0.377}$ eV and $E_2 = \underline{1.51}$ eV

Chapter 5

16. $\int_{-\infty}^{\infty} |\psi|^2 \, dx = 1 \qquad A^2 \int_{-L/4}^{L/4} \cos^2(2\pi x/L) \, dx = 1$

$A^2L/4 = 1 \quad \rightarrow \quad A = 2/(L)^{1/2}$

Probability of finding the particle between 0 and L/8 is

$\int_{0}^{L/8} |\psi|^2 \, dx = A^2 \int_{0}^{L/8} \cos^2(2\pi x/L) \, dx = (1/4) + 1/2\pi = \underline{0.409}$

---

17. Since the wave function for a particle in a one-dimension box of width L is given by $\psi_n = A \sin(n\pi x/L)$, it follows that the probability density is $P(x) = |\psi_n|^2 = A^2 \sin^2(n\pi x/L)$ which is sketched below:

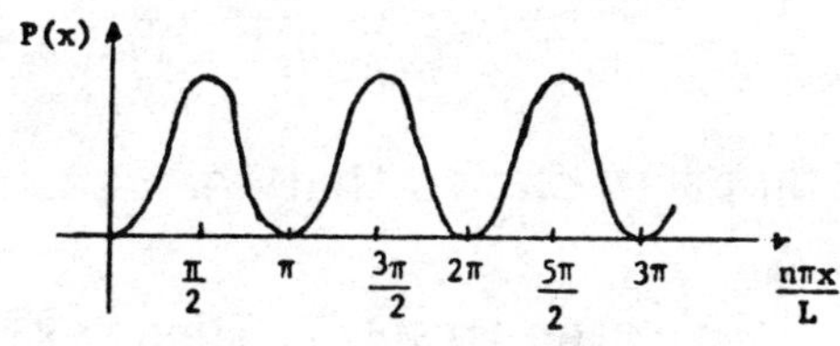

From this sketch we see that P(x) is a <u>maximum</u> when

$n\pi x/L = \pi/2, 3\pi/2, 5\pi/2 \ldots = \pi(m + 1/2)$ or when

$x = (L/n)(m + 1/2) \qquad m = 0,1,2,3 \ldots n$

Likewise, P(x) is a <u>minimum</u> when

$n\pi x/L = 0, \pi, 2\pi, 3\pi \ldots = m\pi$ or when

$x = Lm/n \qquad m = 0,1,2,3 \ldots n$

Chapter 5

18. The possible particle positions within the box are weighted according to the probability density $|\psi|^2 = (2/L)\sin^2(n\pi x/L)$. The average position is calculated as

$$<x> = \int_0^L x|\psi|^2\, dx = (2/L) \int_0^L x \sin^2(\pi x/L)\, dx$$

Making the change of variable $\theta = \pi x/L$ (so that $d\theta = \pi dx/L$) gives

$$<x> = (2L/\pi^2) \int_0^\pi \theta \sin^2 n\theta\, d\theta$$

Using the trigonometric identity $2\sin^2\theta = 1 - \cos 2\theta$, we get

$$<x> = (L/\pi^2)\left\{\int_o^\pi \theta d\theta = \int_o^\pi \theta \cos 2n\theta\, d\theta\right\}$$

An integration by parts shows that the second integral vanishes, while the first integrates to $\pi^2/2$. Thus, $<x> = L/2$, independent of n. For the computation of $<x^2>$, there is an extra factor of x in the integrand. After changing variables to $\theta = \pi x/L$, we get

$$<x^2> = (L^2/\pi^3)\left\{\int_0^\pi \theta^2\, d\theta - \int_0^\pi \theta^2 \cos 2n\theta\, d\theta\right\}$$

The first integral evaluates to $\pi^3/3$; the second may be integrated by parts twice to get

$$\int_0^\pi \theta^2 \cos 2n\theta \, d\theta = -(1/n)\int_0^\pi \theta \sin 2n\theta \, d\theta = (1/2n^2)\theta \cos 2n\theta \Big|_0^\pi = \pi/2n^2$$

Then

$$\langle x^2 \rangle = (L^2/\pi^3)\{\pi^3/3 - \pi/2n^2\} = L^2/3 - L^2/2(n\pi)^2$$

---

19. The allowed energies for this system are given by Eq. 5.18, or

$$E_n = n^2\pi^2\hbar^2/2mL^2 = n^2h^2/8mL^2$$

Using $E_n = 10^{-3}$ J, $m = 10^{-3}$ kg, $L = 10^{-2}$ m, and solving for n gives

$$n = \{8(10^{-3}\text{ kg})(10^{-2}\text{ m})^2)(10^{-3}\text{ J})\}^{1/2}/(6.63 \times 10^{-34}\text{ J}\cdot\text{s})$$

$$n = \underline{4.27 \times 10^{28}}$$

The excitation energy is $\Delta E = E_{n+1} - E_n$, or

$$\Delta E = (h^2/8mL^2)\{(n+1)^2 - n^2\} = (h^2/8mL^2)\{2n+1\}$$

$$= E_n\{(2n+1)/n^2\} \approx (2/n)E_n \qquad \text{for } n >> 1$$

Thus,

$$\Delta E \approx (2)(10^{-3}\text{ J})/(4.27 \times 10^{28}) = \underline{4.69 \times 10^{-32}}\text{ J}$$

---

20. The time development of Ψ is given by Eq. 5.8 or

$$\Psi(x,t) = \int a(k)e^{i\{kx - \omega(k)t\}}dk = (C\alpha/\sqrt{\pi})\int_{-\infty}^{\infty} e^{\{ikx - \omega(k)t - \alpha^2k^2\}}dk,$$

with $\omega(k) = \hbar k^2/2m$ for a free particle of mass m. As in Example 5.3, the integral may be reduced to a recognizable form by completing the square in the exponent. Since $\omega(k)t = (\hbar t/2m)k^2$, we group this term together with α^2k^2 by introducing $\beta^2 = \alpha^2 + \hbar t/2m$ to get

$$ikx - \omega(k)t - \alpha^2k^2 = -(\beta k - ix/2\beta)^2 - x^2/4\beta^2$$

Then, changing variables to $z = \beta k - ix/2\beta$ gives

$$\Psi(x,t) = (C\alpha/\beta\sqrt{\pi})\, e^{-x^2/4\beta^2} \int_{-\infty}^{\infty} e^{-z^2} dz$$

$$= (C\alpha/\beta)\, e^{-x^2/4\beta^2}$$

We see that $\Psi(x,t)$ remains Gaussian in shape, but with a diminishing amplitude $C\alpha/\beta(t) = C\alpha/(\alpha^2 + \hbar t/2m)^{1/2}$, and a width $\Delta x(t) = \beta$ which grows ever larger as

$$\Delta x(t) = (\alpha^2 + \hbar t/2m)^{1/2} = [\{\Delta x(0)\}^2 + \hbar t/2m]^{1/2}.$$

---

21. A particle within the well is subject to no forces and, hence, moves with uniform speed, spending equal time in all parts of the well. Thus, for such a particle the probability density is uniform. That is,

$$P_c(x) = \text{constant}$$

The constant is fixed by requiring the integrated probability to be unity, that is,

$$1 = \int_0^L P_c(x)\, dx = CL$$

or $C = 1/L$.

To find $\langle x\rangle$ we weight the possible particle positions according to the probability density P_c to get

$$\langle x\rangle = \int_0^L x P_c(x)\,dx = (1/L)(x^2/2)\Big|_0^L = L/2$$

Similarly, $\langle x^2\rangle$ is found by weighting the possible values of x^2 with P_c:

$$\langle x^2\rangle = \int_0^L x^2 P_c(x)\,dx = (1/L)(x^3/3)\Big|_0^L = \underline{L^2/3}$$

The classical and quantum results for $\langle x\rangle$ agree exactly; for $\langle x^2\rangle$ the quantum prediction is smaller by an amount $L^2/2(n\pi)^2$ which, however, goes to zero in the limit of large quantum numbers n, where classical and quantum results must coincide (correspondence principle).

---

22. Normalization requires

$$1 = \int_{-\infty}^{\infty} |\psi|^2 dx = C^2\int_0^{\infty} e^{-2x}(1 - e^{-x})^2\,dx = C^2\int_0^{\infty}(e^{-2x} - 2e^{-3x} + e^{-4x})\,dx$$

The integrals are elementary and give

$$1 = C^2\{1/2 - 2(1/3) + 1/4\} = C^2/12$$

The proper units for C are those of $(\text{length})^{-1/2}$; thus, normalization requires $C = (12)^{1/2}\ \text{nm}^{-1/2}$.

Chapter 5

The most likely place for the electron is where the probability $|\psi|^2$ is largest. This is also where ψ itself is largest, and is found by setting the derivative $d\psi/dx$ equal to zero:

$$0 = d\psi/dx = C\{-e^{-x} + 2e^{-2x}\} = Ce^{-x}\{2e^{-x} - 1\}$$

The RHS vanishes when $x = \infty$ (a minimum), and when $2e^{-x} = 1$, or $x = \ln 2$. Thus, the most likely position is at $x_p = \ln 2 = 0.693$ nm.

The average position is calculated from

$$\langle x \rangle = \int_{-\infty}^{\infty} x|\psi|^2\, dx = C^2 \int_0^{\infty} xe^{-2x}(1 - e^{-x})^2\, dx = C^2 \int_0^{\infty} x\{e^{-2x} - 2e^{-3x} + e^{-4x}\}\, dx$$

The integrals are readily evaluated with the help of the formula

$$\int_0^{\infty} xe^{-ax}\, dx = 1/a^2$$

to get

$$\langle x \rangle = C^2\{1/4 - 2(1/9) + 1/16\} = C^2\{13/144\}$$

Substituting $C^2 = 12$ gives $\langle x \rangle = 13/12 = \underline{1.083}$ nm. We see that $\langle x \rangle$ is somewhat greater than the most probable position, since the probability density is skewed in such a way that values of x larger than x_p are weighted more heavily in the calculation of the average.

Chapter 5

23. (a) Not acceptable -- diverges as $x \to \infty$.
    (b) Acceptable.
    (c) Acceptable.
    (d) Not acceptable -- not a single-valued function.
    (e) Not acceptable -- the wave is discontinuous (as is the slope).

---

24. The Schrodinger equation, after rearrangement, is

$$d^2\psi/dx^2 = (2m/\hbar^2)\{U(x) - E\}\,\psi(x)$$

In the well interior, $U(x) = 0$ and solutions to this equation are $\sin kx$ and $\cos kx$, where $k^2 = 2mE/\hbar^2$. The waves symmetric about the midpoint of the well ($x = 0$) are described by

$$\psi(x) = A\cos kx \qquad -L < x < +L$$

In the region outside the well, $U(x) = U$, and the independent solutions to the wave equation are $e^{\pm\alpha x}$, with $\alpha^2 = (2m/\hbar^2)(U - E)$. The growing exponentials must be discarded to keep the wave from diverging at infinity. Thus, the waves in the exterior region which are symmetric about the midpoint of the well are given by

$$\psi(x) = C\,e^{-\alpha|x|} \qquad x > L \quad \text{or} \quad x < -L$$

At $x = L$, continuity of ψ requires

$$A\cos kL = C\,e^{-\alpha L}$$

For the slope to be continuous here, we also must require

$$-Ak\sin kL = -C\alpha\,e^{-\alpha L}$$

Dividing the two equations gives the desired restriction on the allowed energies:

$$k \tan kL = \alpha$$

The dependence on E (or k) is made more explicit by noting that

$k^2 + \alpha^2 = 2mU/\hbar^2$, which allows the energy condition to be written

$$k \tan kL = \{(2mU/\hbar^2) - k^2\}^{1/2}$$

Multiplying by L, squaring the result, and using $\tan^2\theta + 1 = \sec^2\theta$ gives

$$(kL)^2 \sec^2(kL) = 2mUL^2/\hbar^2$$

from which the desired form follows immediately.

The ground state is the symmetric waveform having the lowest energy. For electrons in a well of height U = 5 eV and width 2L = 0.2 nm, we calculate

$$2mUL^2/\hbar^2 = (2)(511 \times 10^3 \text{ eV}/c^2)(5 \text{ eV})(0.1 \text{ nm})^2/(197.3 \text{ eV}\cdot\text{nm}/c)^2$$

$$= 1.3127$$

With this value, the equation for $\theta = kL$

$$\theta/\cos\theta = (1.3127)^{1/2} = 1.1457$$

can be solved numerically employing methods of varying sophistication. The simplest of these is trial and error, which gives $\theta = 0.799$. From this, we find $k = 7.99 \text{ nm}^{-1}$, and an energy

$$E = \hbar^2k^2/2m = (197.3 \text{ eV}\cdot\text{nm}/c)^2(7.99 \text{ nm}^{-1})^2/\{(2)(511 \times 10^3 \text{ eV}/c^2)\}$$

$= \underline{2.432}$ eV.

Chapter 5

25. Inside the well, the particle is free and the Schrodinger waveform is trigonometric with wavenumber $k = (2mE/\hbar^2)^{1/2}$:

$$\psi(x) = A \sin kx + B \cos kx \qquad 0 \leq x \leq L$$

The infinite wall at $x = 0$ requires $\psi(0) = B = 0$.

Beyond $x = L$, $U(x) = U$ and the Schrodinger equation

$$d^2\psi/dx^2 = (2m/\hbar^2)\{U - E\}\psi(x)$$

which has exponential solutions for $E < U$

$$\psi(x) = Ce^{-\alpha x} + De^{+\alpha x}, \qquad x > L$$

where $\alpha = [2m(U - E)/\hbar^2]^{1/2}$. To keep ψ bounded at $x = \infty$, we must take $D = 0$. At $x = L$, continuity of ψ and $d\psi/dx$ demands

$$A \sin kL = C e^{-\alpha L}$$

$$kA \cos kL = -\alpha C e^{-\alpha L}$$

Dividing one by the other gives an equation for the allowed particle energies:

$$k \cot kL = -\alpha$$

The dependence on E (or k) is made more explicit by noting that $k^2 + \alpha^2 = 2mU/\hbar^2$, which allows the energy condition to be written

$$k \cot kL = -[(2mU/\hbar^2) - k^2]^{1/2}$$

Multiplying by L, squaring the result, and using $\cot^2\theta + 1 = \csc^2\theta$ gives

$$(kL)^2 \csc^2 (kL) = 2mUL^2/\hbar^2$$

from which we obtain

$$kL/\sin kL = (2mUL^2/\hbar^2)^{1/2}$$

Since $\theta/\sin\theta$ is never smaller than unity for <u>any</u> value of θ, there can be no bound state energies if

$$2mUL^2/\hbar^2 < 1$$

---

26. The symmetry of $|\psi(x)|^2$ about $x = 0$ can be exploited effectively in the calculation of average values. To find $\langle x \rangle$

$$\langle x \rangle = \int_{-\infty}^{\infty} x|\psi(x)|^2 dx$$

we notice that the integrand is antisymmetric about $x = 0$ due to the extra factor of x (an odd function). Thus, the contribution from the two half-axes $x > 0$ and $x < 0$ cancel exactly, leaving $\langle x \rangle = 0$. For the calculation of $\langle x^2 \rangle$, however, the integrand is symmetric and the half-axes contribute equally to the value of the integral, giving

$$\langle x^2 \rangle = 2\int_0^{\infty} x^2|\psi(x)|^2 dx = 2C^2 \int_0^{\infty} x^2 e^{-2x/x_0} dx$$

Two integrations by parts show the value of the integral to be $2\,(x_0/2)^3$. Upon substituting for C^2, we get

$$\langle x^2 \rangle = 2(1/x_0)(2)(x_0/2)^3 = x_0^2/2$$

and

$$\Delta x = (\langle x^2\rangle - \langle x\rangle^2)^{1/2} = (x_o^2/2)^{1/2} = x_o/\sqrt{2}$$

In calculating the probability for the interval $-\Delta x$ to $+\Delta x$, we appeal to symmetry once again to write

$$P = \int_{-\Delta x}^{+\Delta x} |\psi|^2\,dx = 2C^2\int_0^{\Delta x} e^{-2x/x_o}\,dx = -2C^2(x_o/2)e^{-2x/x_o}\Big|_0^{\Delta x}$$

$$= 1 - e^{-\sqrt{2}} = \underline{0.757}$$

or about 75.7%, independent of x_o.

---

27. Applying the momentum operator $[p] = (\hbar/i)d/dx$ to each of the candidate functions yields

(a) $[p]\{A\sin(kx)\} = (\hbar/i)k\{A\cos(kx)\}$

(b) $[p]\{A\sin(kx) - A\cos(kx)\} = (\hbar/i)k\{A\cos(kx) + A\sin(kx)\}$

(c) $[p]\{A\cos(kx) + iA\sin(kx)\} = (\hbar/i)k\{-A\sin(kx) + iA\cos(kx)\}$

(d) $[p]\{e^{ik(x-a)}\} = (\hbar/i)ik\{e^{ik(x-a)}\}$

In the case (c), the result is a multiple of the original function, since

$$-A\sin(kx) + iA\cos(kx) = i\{A\cos(kx) + iA\sin(kx)\}$$

The multiple is $(\hbar/i)(ik) = \hbar k$, and is the eigenvalue. Likewise for (d), the operation $[p]$ returns the original function with the multiplier $\hbar k$.

Thus, (c) and (d) are eigenfunctions of [p] with eigenvalue $\hbar k$, while (a) and (b) are not eigenfunctions of this operator.

---

28. The probability density for this case is $|\psi_0(x)|^2 = C_0^2 e^{-ax^2}$, with $C_0 = (a/\pi)^{1/4}$ and $a = m\omega/\hbar$. For the calculation of the average position

$$\langle x \rangle = \int_{-\infty}^{\infty} x |\psi_0(x)|^2 \, dx$$

we note that the integrand is an odd function, so that the integral over the negative half-axis $x < 0$ exactly cancels that over the positive half-axis ($x > 0$), leaving $\langle x \rangle = 0$. For the calcuation of $\langle x^2 \rangle$, however, the integrand $x^2 |\psi_0|^2$ is symmetric, and the two half-axes contribute equally, giving

$$\langle x^2 \rangle = 2C_0^2 \int_0^{\infty} x^2 e^{-ax^2} dx = 2C_0^2 (1/4a)(\pi/a)^{1/2}$$

Substituting for C_0 and a give $\langle x^2 \rangle = 1/2a = \hbar/2m\omega$ and

$$\Delta x = (\langle x^2 \rangle - \langle x \rangle^2)^{1/2} = (\hbar/2m\omega)^{1/2}$$

Chapter 5

29. Since there is no preference for motion in the leftward sense vs. the rightward sense, a particle would spend equal time moving left as moving right, suggesting $\langle p \rangle = 0$.

To find $\langle p^2 \rangle$, we express the average energy as the sum of its kinetic and potential energy contributions:

$$\langle E \rangle = \langle p^2/2m \rangle + \langle U \rangle = \langle p^2 \rangle/2m + \langle U \rangle$$

But energy is sharp in the oscillator ground state, so that $\langle E \rangle = E_0 = (1/2)\hbar\omega$. Furthermore, remembering that $U(x) = (1/2)m\omega^2x^2$ for the quantum oscillator, and using $\langle x^2 \rangle = \hbar/2m\omega$ from Problem 5.28, gives

$$\langle U \rangle = (1/2)m\omega^2\langle x^2 \rangle = (1/4)\hbar\omega$$

Then

$$\langle p^2 \rangle = 2m(E_0 - \langle U \rangle) = 2m(\hbar\omega/4) = m\hbar\omega/2$$

and

$$\Delta p = (\langle p^2 \rangle - \langle p \rangle^2)^{1/2} = (m\hbar\omega/2)^{1/2}$$

---

30. From Problems 28 and 29, we have $\Delta x = (\hbar/2m\omega)^{1/2}$ and $\Delta p = (m\hbar\omega/2)^{1/2}$. Thus,

$$\Delta x \, \Delta p = (\hbar/2m\omega)^{1/2}(m\hbar\omega/2)^{1/2} = \hbar/2$$

for the oscillator ground state. This is the minimum uncertainty product permitted by the uncertainty principle, and is realized only for the ground state of the quantum oscillator.

Chapter 5

31. After rearrangement, the Schrodinger equation is

$$d^2\psi/dx^2 = (2m/\hbar^2)\{U(x) - E\}\,\psi(x)$$

with $U(x) = (1/2)m\omega^2x^2$ for the quantum oscillator. Differentiating $\psi(x) = C\,x\,e^{-\alpha x^2}$ gives

$$d\psi/dx = -2\alpha x\,\psi(x) + C\,e^{-\alpha x^2}$$

and

$$d^2\psi/dx^2 = -2\alpha x\,d\psi/dx - 2\alpha\,\psi(x) - (2\alpha x)C\,e^{-\alpha x^2}$$

$$= (2\alpha x)^2\psi(x) - 6\alpha\psi(x)$$

Therefore, for $\psi(x)$ to be a solution requires

$$(2\alpha x)^2 - 6\alpha = (2m/\hbar^2)\{U(x) - E\} = (m\omega/\hbar)^2x^2 - 2mE/\hbar^2$$

Equating coefficients of like terms gives

$$2\alpha = m\omega/\hbar \qquad \text{and} \qquad 6\alpha = 2mE/\hbar^2$$

Thus, $\alpha = m\omega/2\hbar$ and $E = 3\alpha\hbar^2/m = (3/2)\hbar\omega$.

The normalization integral is

$$1 = \int_{-\infty}^{\infty} |\psi(x)|^2\,dx = 2C^2\int_{0}^{\infty} x^2e^{-2\alpha x^2}\,dx$$

where the second step follows from the symmetry of the integrand about $x = 0$. Identifying a with 2α in the formula of Problem 28 gives

$$1 = 2C^2(1/8\alpha)(\pi/2\alpha)^{1/2} \qquad \text{or} \qquad C = (32\alpha^3/\pi)^{1/4}$$

Chapter 5

32. At its limits of vibration $x = \pm A$, the classical oscillator has all its energy in potential form:

$$E = (1/2)m\omega^2 A^2 \qquad \text{or } A = (2E/m\omega^2)^{1/2}$$

If the energy is quantized as $E_n = (n + 1/2)\hbar\omega$, then the corresponding amplitudes are

$$A_n = [(2n + 1)\hbar/m\omega]^{1/2}$$

---

33. $P_c(x)\,dx$ is proportional to the time which the particle spends in the interval dx. This time dt is inversely related to its speed v as $dt = dx/v$, so that $P_c(x)\,dx = C\,dt$, or $P_c(x) = C/v$. But the speed of the oscillator varies with its position in such a way as to keep the total energy constant:

$$E = (1/2)mv^2 + (1/2)m^2x^2 \qquad \text{or} \qquad v^2 = 2E/m - \omega^2 x^2$$

Writing E in terms of the classical amplitude as $E = (1/2)m\omega^2 A^2$ gives

$$v = \omega(A^2 - x^2)^{1/2}$$

and

$$P_c(x) = (C/\omega)(A^2 - x^2)^{-1/2}$$

The constant C is a normalizing factor chosen to ensure a total probability of one:

$$1 = \int_{-A}^{A} P_c(x)\,dx = (C/\omega)\int_{-A}^{A} (A^2 - x^2)^{-1/2}\,dx$$

The integral is evaluated with the trigonometric substitution $x = A \sin\theta$ (so that $dx = A \cos\theta\, d\theta$) to get

$$1 = (C/\omega)\int_{-\pi/2}^{\pi/2} d\theta = \pi C/\omega$$

Thus, C/ω is just $1/\pi$ and

$$P_c(x) = (1/\pi)/(A^2 - x^2)^{1/2}$$

for a classical oscillator with amplitude of vibration equal to A.

---

34. Normalization requires

$$1 = \int_{-\infty}^{\infty} |\Psi|^2\,dx = C^2\int_{-\infty}^{\infty} \{\psi_1^* + \psi_2^*\}\{\psi_1 + \psi_2\}\,dx$$

$$= C^2 \left\{\int |\psi_1|^2\,dx + \int |\psi_2|^2\,dx + \int \psi_2^*\psi_1\,dx + \int \psi_1^*\psi_2\,dx\right\}$$

The first two integrals on the right are unity, while the last two are, in fact, the same integral since ψ_1 and ψ_2 are both real. Using the waveforms for the infinite square well, we find

$$\int \psi_2\psi_1\,dx = (2/L)\int_0^L \sin(\pi x/L)\sin(2\pi x/L)\,dx$$

$$= (1/L)\int_0^L \{\cos(\pi x/L) - \cos(3\pi x/L)\}\,dx$$

where, in writing the last line, we have used the trigonometric identities $\cos(A \pm B) = \cos(A)\cos(B) - \sin(A)\sin(B)$. Both of the integrals remaining are readily evaluated, and are zero. Thus,

$$1 = C^2\{1 + 1 + 0 + 0\} = 2C^2, \qquad \text{or} \qquad C = 1/\sqrt{2}$$

Since $\psi_{1,2}$ are stationary states, they develop in time according to their respective energies $E_{1,2}$ as $e^{-iEt/\hbar}$. Then

$$\Psi(x,t) = C\,\{\psi_1\, e^{-iE_1t/\hbar} + \psi_2\, e^{-iE_2t/\hbar}\}$$

$\Psi(x,t)$ is a stationary state only if it is an eigenfunction of the energy operator $[E] = i\hbar\partial/\partial t$. Applying $[E]$ to Ψ gives

$$[E]\Psi = C\,\{(i\hbar(-iE_1/\hbar)\psi_1\, e^{-iE_1t/\hbar} + i\hbar(-iE_2/\hbar)\psi_2\, e^{-iE_2t/\hbar}\}$$

$$= C\,\{E_1\psi_1\, e^{-iE_1t/\hbar} + E_2\psi_2\, e^{-iE_2t/\hbar}\}$$

Since $E_1 \neq E_2$, the operation $[E]$ does <u>not</u> return a multiple of the wavefunction, and so Ψ is not a stationary state. Nonetheless, we may calculate the average energy for this state as

$<E> = \int \Psi^*[E]\Psi \, dx$

$$= C^2 \int \{\psi_1^* e^{+iE_1t/\hbar} + \psi_2^* e^{+iE_2t/\hbar}\}\{E_1\psi_1 e^{-iE_1t/\hbar} + E_2\psi_2 e^{-iE_2t/\hbar}\}dx$$

$$= C^2 \{E_1 \int |\psi_1|^2 dx + E_2 \int |\psi_2|^2 dx\}$$

Since $\psi_{1,2}$ are normalized and $C^2 = 1/2$, we get finally

$$<E> = (E_1 + E_2)/2$$

---

35. The average position at any instant is given by

$$<x> = \int_{-\infty}^{\infty} x|\Psi|^2 dx$$

$$= C^2 \int_{-\infty}^{\infty} x\{\psi_1^* e^{+iE_1t/\hbar} + \psi_2^* e^{+iE_2t/\hbar}\}\{\psi_1 e^{-iE_1t/\hbar} + \psi_2 e^{-iE_2t/\hbar}\}dx$$

$$= C^2 \{\int x|\psi_1|^2dx + \int x|\psi_2|^2dx + e^{-i\Omega t}\int x\psi_1^*\psi_2 dx + e^{+i\Omega t}\int x\psi_2^*\psi_1 dx\}$$

where $\Omega = (E_2 - E_1)/\hbar$. The last two integrals on the right are identical, since $\psi_{1,2}$ are real. Furthermore, $e^{-i\Omega t} + e^{+i\Omega t} = \cos(\Omega t)$, and $C^2 = 1/2$ from Problem 35. Thus, the result takes the form

$$<x> = x_0 + A\cos(\Omega t)$$

with the definitions given.

To evaluate x_0, we note that $<x> = L/2$ for any stationary state of the well. Therefore, $x_0 = (1/2)\{L/2 + L/2\} = L/2 = 0.5$ nm, no matter which two stationary states we use in the superposition. To find A, we use the ground and first excited state waves of the infinite well to write

$$A = (2/L)\int_0^L x\sin(\pi x/L)\sin(2\pi x/L)\,dx$$

$$= (1/L)\int_0^L x\{\cos(\pi x/L) - \cos(3\pi x/L)\}\,dx$$

Integrating by parts once, we obtain

$$A = (1/L)(L/3\pi)x\{3\sin(\pi x/L) - \sin(3\pi x/L)\}\Big|_0^L$$

$$- (1/L)(L/3\pi)\int_0^L \{3\sin(\pi x/L) - \sin(3\pi x/L)\}\,dx$$

$$= 0 + (1/L)(L/3\pi)^2\{9\cos(\pi x/L) - \cos(3\pi x/L)\}\Big|_0^L$$

$$= (L/9\pi^2)\{-9 + 1 - 9 + 1\} = -16L/9\pi^2 = -0.18 \text{ nm}.$$

For electrons in this well we have the energie

$$E_1 = h^2/8mL^2 = (1.24 \text{ keV·nm}/c)^2/[8(511 \text{ keV}/c^2)(1 \text{ nm})^2]$$

$$= 0.376 \text{ eV}$$

and

$$E_2 = (2)^2E_1 = 1.50 \text{ eV}.$$

Chapter 5

The period of oscillation is $T = 2\pi/\Omega$, or

$$T = \hbar/(E_2 - E_1) = (4.136 \times 10^{-15}\ \text{eV}\cdot\text{s})/(1.128\ \text{eV}) = 3.66 \times 10^{-15}\ \text{s}$$

A classical electron with (kinetic) energy $(E_1 + E_2)/2 = 0.94$ eV

would have speed

$$v = (2E/m)^{1/2} = (1.92 \times 10^{-3})c$$

and would require $2L/v = 3.47 \times 10^{-15}$ s to shuttle back and forth in

the well one time, a distance $2L = 2$ nm.

Chapter 6

1. The reflection coefficient is the ratio of the reflected to the incident wave intensity, or

$$R = |(1/2)(1 - i)|^2/|(1/2)(1 + i)|^2$$

But $|1 - i|^2 = (1 - i)(1 - i)^* = (1 - i)(1 + i) = |1 + i|^2$, so that $R = 1$ in this case.

To the left of the step, the particle is free and the solutions to Schrodinger's equation are $e^{\pm ikx}$ with wavenumber $k = (2mE/\hbar^2)^{1/2}$. To the right of the step, $U(x) = U$ and the wave equation is

$$d^2\psi/dx^2 = (2m/\hbar^2)\{U - E\}\,\psi(x)$$

With $\psi(x) = e^{-kx}$, we find $d^2\psi/dx^2 = k^2\psi(x)$, so that

$$k = [2m(U - E)/\hbar^2]^{1/2}$$

Substituting $k = (2mE/\hbar^2)^{1/2}$ shows that

$$[E/(U - E)]^{1/2} = 1 \qquad \text{or} \qquad E/U = 1/2$$

For 10 MeV protons, $E = 10$ MeV and $m = 938.28\ \text{MeV}/c^2$. Using $\hbar = 197.3$ MeV·fm (1 fm $= 10^{-15}$ m), we find

$$\delta = 1/k = \hbar/(2mE)^{1/2}$$

$$= (197.3\ \text{MeV·fm})/[(2)(938.28\ \text{MeV}/c^2)(10\ \text{MeV})]^{1/2} = \underline{1.44}\ \text{fm}$$

---

2. To the left of the step the particle is free with kinetic energy E and corresponding wavenumber $k_1 = (2mE/\hbar^2)^{1/2}$:

$$\psi(x) = A\,e^{ik_1x} + B\,e^{-ik_1x} \qquad x \le 0$$

To the right of the step the kinetic energy is reduced to $E - U$ and the wavenumber is now $k_2 = [2m(E - U)/\hbar^2]^{1/2}$:

$$\psi(x) = C\,e^{ik_2x} + D\,e^{-ik_2x} \qquad x \ge 0$$

with $D = 0$ for waves incident on the step from the left. At $x = 0$

both ψ and $d\psi/dx$ must be continuous:

$$\psi(0) = A + B = C$$

$$d\psi/dx\big|_0 = ik_1(A - B) = ik_2C$$

Eliminating C gives

$$A + B = (k_1/k_2)(A - B) \qquad \text{or} \qquad A(k_1/k_2 - 1) = B(k_1/k_2 + 1)$$

Thus,

$$R = |B/A|^2 = (k_1/k_2 - 1)^2/(k_2 + 1)^2 = (k_1 - k_2)^2/(k_1 + k_2)^2$$

As $E \to U$, $k_2 \to 0$ and $R \to 1$ (no transmission), in agreement with the result for any energy $E < U$. For $E \to \infty$, $k_1 \to k_2$ and $R \to 0$ (perfect transmission), suggesting correctly that very energetic particles do not 'see' the step and so are unaffected by it.

---

3. With E = 25 MeV and U = 20 MeV, the ratio of wavenumbers is

$$k_1/k_2 = [E/(E - U)]^{1/2} = [25/(25 - 20)]^{1/2} = \sqrt{5} = 2.236$$

Then, from Problem 2

$$R = (\sqrt{5} - 1)^2/(\sqrt{5} + 1)^2 = 0.146$$

and

$$T = 1 - R = 0.854$$

Thus, 14.6% of the incoming particles would be reflected and 85.4% transmitted. For electrons with the <u>same</u> energy, the transparency and reflectivity of the step are unchanged.

---

4. The reflection coefficient for this case is given in Problem 2 as

$$R = |B/A|^2 = (k_1/k_2 - 1)^2/(k_1/k_2 + 1)^2 = (K_1 - k_2)^2/(k_1 + k_2)^2$$

The wavenumbers are those for electrons with kinetic energies $E = 54$ eV and $E - U = 54 + 10 = 64$ eV:

$$k_1/k_2 = [E/(E - U)]^{1/2} = [54/(64)]^{1/2} = 0.9186$$

Then

$$R = (0.9186 - 1)^2/(0.9186 + 1)^2 = 1.80 \times 10^{-3}$$

is the fraction of the incident beam which is reflected at the boundary.

---

5. Let P_T = (prob. of finding the particle to right of barrier)

÷ (prob. of finding the particle to left of barrier)

P_T is approximately equal to $|\Psi_{III}|^2/|\Psi_I|^2 = (e^{-CL})^2 = e^{-2CL}$

Numerical estimates

1) For $m = 9.11 \times 10^{-31}$ kg, $U - E = 1.6 \times 10^{-21}$ J, $L = 10^{-10}$ m;

$C = [2m(U - E)]^{1/2}/h = 5.12 \times 10^8\ \text{m}^{-1}$

$P_T = e^{-2CL} = \ = \underline{0.90}$

2) For $m = 9.11 \times 10^{-31}$ kg, $U - E = 1.6 \times 10^{-18}$ J, $L = 10^{-10}$ m;

$C = 5.12 \times 10^9\ \text{m}^{-1}$;

$P_T = e^{-1.02} = \underline{0.36}$

3) For $m = 6.7 \times 10^{-27}$ kg, $U - E = 1.6 \times 10^{-13}$ J, $L = 10^{-15}$ m;

$C = 4.4 \times 10^{14}$ m;

$P_T = e^{-0.888} = \underline{0.41}$

4) For $m = 8$ kg, $U - E = 1$ J, $L = 0.02$ m, $C = 3.8 \times 10^{34}$;

$P_T = e^{-2CL} = e^{-1.5 \times 10^{33}} = 10^{-6.6 \times 10^{32}} = 0$

---

6. For a wide barrier, we have

$$T(E) = \approx [4k\delta/(1 + (k\delta)^2)]^2 e^{-2L/\delta}$$

For 0.1% transmission $T(E) = 0.001$, and the resulting equation must be solved for E using $L = 2$ nm and $U = 5$ eV. We adopt $x = E/U$ as the unknown and write

$$\delta = \hbar/[2m(U - E)]^{1/2} = (\hbar/[2mU]^{1/2})(1 - x)^{-1/2}$$

$$2L/\delta = (1 - x)^{1/2}\, 2(2\text{ nm})[2(511 \times 10^3\text{ eV}/c^2)(5\text{ eV})]^{1/2}/(197.3\text{ eV}\cdot\text{nm}/c)$$

$$= 45.829\,(1 - x)^{1/2} \quad \text{and}$$

$$(k\delta)^2 = E/(U - E) = x/(1 - x)$$

The equation for x is then

$$0.001 = 16x(1 - x)e^{-45.829\,(1 - x)^{1/2}}$$

The solution is most easily obtained by trial and error. Using this method we find $x = 0.9862$, implying

$$E = xU = (0.9862)(5\text{ eV}) = 4.931\text{ eV}$$

Finally, we compute

$$2L/\delta = 45.829\,(1 - x)^{1/2} = 5.38 \qquad \text{or} \qquad L/\delta = 2.69$$

In this case, L exceeds δ by only a factor of 2, and the use of the wide barrier result may be questioned, particularly if great accuracy is desired.

---

7. The wide barrier approximation is valid if $L \gg \delta$, but δ is never less than its value for zero energy, so a necessary criterion is

$$L \gg \delta\big|_{E=0} = \hbar/(2mU)^{1/2}$$

After squaring both sides and rearranging, we get

$$2mUL^2/\hbar^2 \gg 1$$

For electrons impinging on a barrier 5 eV high and 2 nm wide, the left hand side is

$$2(511 \times 10^3\text{ eV}/c^2)(5\text{ eV})(2\text{ nm})^2/(197.3\text{ eV}\cdot\text{nm}/c)^2 = 525$$

which certainly is much larger than 1. Thus, at least for energies not too near the top of the barrier, the wide barrier result should apply.

8. For $E > 0$ solutions to the wave equation on either side of the origin are free particle plane waves with wavenumber $k = (2mE/\hbar^2)^{1/2}$:

$$\psi(x) = Ae^{+ikx} + Be^{-ikx} \qquad \text{for } x < 0$$

$$\psi(x) = Fe^{+ikx} + Ge^{-ikx} \qquad \text{for } x > 0$$

We take $G = 0$ (no refected wave in the region to the right of the well) for particles incident on the delta well from the left. Some fraction of these are transmitted, as given by $T = |F/A|^2$. To find T we impose the slope condition on the waveform to get

$$ik(A - B) - ikF = -(2mS/\hbar^2)F$$

and demand continuity of the wave at $x = 0$:

$$A + B = F$$

Dividing the first equation through by ik and adding the result to the second gives

$$A = F\{1 + i(mS/\hbar^2k)\} = F\{1 + i/(-E_0/E)\}$$

In the second step we have written $E = h^2k^2/2m$ and $E_0 = -mS^2/2\hbar^2$ as convenient parametrizations for the scattering problem. The transmission coefficient is

$$T(E) = |A/F|^{-2} = \{1 + (-E_0/E)\}^{-1}$$

The transmission coefficient for the delta well is sketched in the Figure below. T(E) increases with E, approaching 1 (perfect transmission) only asymptotically as E becomes large (since $E_0 < 0$). Although the interpretation of T as a transmission factor is sensible only for nonnegative particle energies, it is interesting that the right hand side becomes infinite if we take

$$E = E_0 = -mS^2/2\hbar^2$$

For $E = |E_0|$, we find $T = 1/2$, so that exactly half of the particles incident on the well with this energy are transmitted, and the other half reflected.

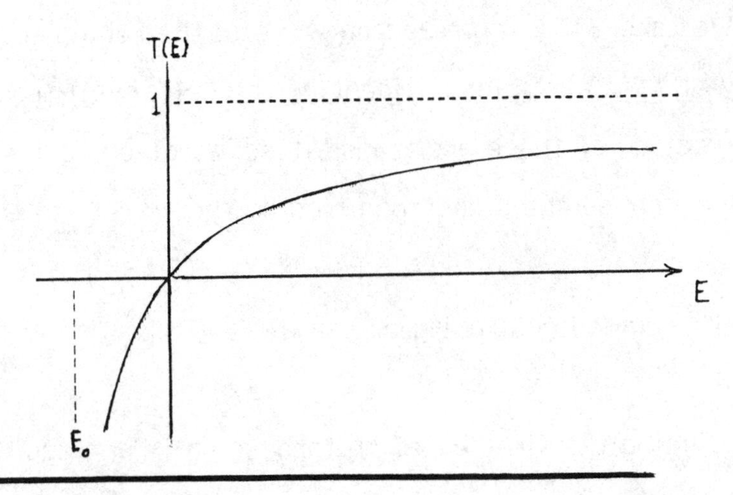

9. As in Problem 8, waveform continuity and the slope condition at the site of the delta well demand

$$A + B = F \quad \text{and} \quad ik(A - B) - ikF = -(2mS/\hbar^2)F$$

Dividing the second of these equations by ik and subtracting from the first gives

$$2B + F = F + (2mS/\hbar^2)F/ik, \quad \text{or}$$

$$B = -i(mS/\hbar^2 k)F = -iF(-E_0/E)^{1/2}$$

Thus, the reflection coefficient R is

$$R(E) = |B/A|^2 = |B/F|^2|F/A|^2 = (-E_0/E)\{1 + (-E_0/E)\}^{-1}$$

Then, with T(E) from Problem 8, we find

$$R(E) + T(E) = (1 - E_0/E)\{1 + (-E_0/E)\}^{-1} = 1$$

10. Since the alpha particle has the combined mass of 2 protons and 2 neutrons, or about 3755.8 MeV/c^2, the first approximation to the decay length δ is

$$\delta \approx \hbar/(2mU)^{1/2} = (197.3 \text{ MeV-fm/c})/\{2(3755.8 \text{ MeV/c}^2)(30 \text{ MeV})\}^{1/2}$$

$$= 0.4156 \text{ fm}$$

This gives an effective width for the (infinite) well of

$R + \delta = 9.4156$ fm, and a ground state energy

$$E_1 = \pi^2(197.3 \text{ MeV-fm/c})^2/\{2(3755.8 \text{ MeV/c}^2)(9.4156 \text{ fm})^2\}$$

$$= 0.577 \text{ MeV}$$

From this E we calculate $U - E = 29.42$ MeV and a new decay length

$$\delta \approx (197.3 \text{ MeV-fm/c})/\{2(3755.8 \text{ MeV/c}^2)(29.42 \text{ MeV})\}^{1/2}$$

$$= 0.4197 \text{ fm}$$

This, in turn, increases the effective well width to 9.4197 fm and lowers the ground state energy to $E_1 = 0.576$ MeV. Since our estimate for E has changed by only 0.001 MeV, we may be content with this value.

With a kinetic energy of E_1, the alpha particle in the ground state has speed

$$v_1 = (2E_1/m)^{1/2} = [2(0.576 \text{ MeV})/(3755.8 \text{ MeV/c}^2)]^{1/2} = 0.0175c$$

In order to be ejected with a kinetic energy of 4.05 MeV, the alpha particle must have been preformed in an excited state of the nuclear well, not the ground state.

---

11. The collision frequency f is the reciprocal of the transit time for the alpha particle crossing the nucleus, or $f = v/2R$ where v is the speed of the alpha. Now v is found from the kinetic energy which, inside the nucleus, is not the total energy E but the difference $E - U$ between the total energy and the potential energy representing the bottom of the nuclear well. At the nuclear radius $R = 9$ fm, the Coulomb energy is

$$k(Ze)(2e)/R = 2Z(ke^2/a_0)(a_0/R) = 2(88)(27.2\text{ eV})(5.29 \times 10^4\text{ fm}/9\text{ fm})$$

$$= 28.14\text{ MeV}$$

From this we conclude that $U = -1.86$ MeV to give a nuclear barrier of 30 MeV overall. Thus an alpha with $E = 4.05$ MeV has kinetic energy $4.05 + 1.86 = 5.91$ MeV inside the nucleus. Since the alpha particle has the combined mass of 2 protons and 2 neutrons, or about 3755.8 MeV/c^2, this kinetic energy represents a speed

$$v = (E_k/m)^{1/2} = [2(5.91)/3755.8\text{ MeV}/c^2]^{1/2} = 0.056c$$

Thus, we find for the collision frequency

$$f = v/2R = (0.056\text{ c})/[2(9\text{ fm})] = \underline{9.35 \times 10^{20}}\text{ Hz}$$

Chapter 6

12. The classical turning points are found from $U(x) = E$, or

$$|x| = a \pm (\hbar/M\omega)^{1/2}$$

There are four roots symmetrically placed about the origin; the limits for the tunneling integral are the two closest to $x = 0$ at $x = \pm x_o$, where

$$x_o = a - (\hbar/M\omega)^{1/2}$$

The tunneling integral itself is

$$I = (2M/\hbar^2)^{1/2} \int [U(x) - E]^{1/2}\, dx$$

$$= M\omega/\hbar \int_{-x_o}^{x_o} [(|x| - a)^2 - \hbar/M\omega]^{1/2}\, dx$$

$$= 2M\omega/\hbar \int_{o}^{x_o} [(x - a)^2 - \hbar/M\omega]^{1/2}\, dx$$

Letting $x - a = -(h/M\omega)^{1/2} \cosh(y)$ $(dx = -(h/M\omega)^{1/2} \sinh(y))$ gives

$$I = 2 \int_{o}^{y_o} \sinh^2(y)dy = \int_{o}^{y_o} [\cosh(2y) - 1]dy = (1/2)\sinh(2y_o) - y_o$$

where $\cosh(y_o) = a(M\omega/\hbar)^{1/2}$

Using $M = 14$ u ($= 2.324 \times 10^{-26}$ kg) for the N atom and the given values for ω and a, we find

$$(\hbar/M\omega)^{1/2} = \{(1.055 \times 10^{-34}\ \text{J-s})/\ [(2.324 \times 10^{-26}\ \text{kg})(10^{14}\ \text{s}^{-1})]\}^{1/2}$$

$$= 0.67 \times 10^{-11}\ \text{m} = 0.067\ \text{Å},$$

and

$$\cosh(y_0) = 0.37/0.067 = 5.49$$

leading to $y_0 = 2.388$ and $I = 27.27$. The tunneling probability is then

$$T = e^{-27.27} = 1.43 \times 10^{-12}$$

For the collision frequency f we use the classical frequency of vibration, or $f = \omega/2\pi = 1.58 \times 10^{13}$ Hz. Then the tunneling rate is

$$\lambda = f \cdot T = (1.59 \times 10^{13}\ \text{Hz})(1.43 \times 10^{-12}) = 22.8\ \text{s}^{-1}$$

which leads to a tunneling time

$$t = 1/\lambda = \underline{4.39 \times 10^{-2}}\ \text{s}$$

---

13. Any one conduction electron of the metal is virtually free to move about with a speed v fixed by its kinetic energy $E_k = (1/2)mv^2$, but the average energy per electron available for motion in any specific direction (say, normal to the surface) is reduced from this by the factor 1/3 to account for the random directions of travel:

$$\langle E_k \rangle = (1/2)m\{v_x^2\rangle + \langle v_y^2\rangle + \langle v_z^2\rangle\} = (3/2)m\langle v_x^2\rangle, \text{ or}$$

$$(1/2)m\langle v_x^2\rangle = (1/3)\langle E_k\rangle$$

For a sample with dimension L normal to the surface, the time elapsed between collisions with this surface is $2L/|v_x|$, for any one electron. The reciprocal of this time is the collision frequency. For two electrons, collisions occur twice as often, and so forth, so that the collision frequency for N electrons is $N|v_x|/2L$. Making the identification $|v_x|^2 = \langle v_x^2 \rangle$ allows us to write the collision frequency f in terms of electron energy as

$$f = (N/2L)(2E_k/3m)^{1/2}$$

The density of copper is 8.96 g/cm^3, so one cubic centimeter represents an amount of copper equal to 8.96 g, or the equivalent of 8.96/63.54 = 0.141 moles (the atomic weight of copper is 63.54). Since each mole contains a number of atoms equal to Avogadro's number $N_A = 6.02 \times 10^{23}$, the number of copper atoms in our sample is $0.141N_A$ or about 8.49×10^{22}, which is also the number N of conduction electrons.

The most energetic electrons in copper have kinetic energies of about 7 eV. Using this for E_k, L = 1 cm, and N = 8.49×10^{22} gives for the collision frequency

$$f = \underline{3.85 \times 10^{30}} \text{ Hz}$$

Chapter 7

1. $E = [\hbar^2\pi^2/2m][(n_x/L_x)^2 + (n_y/L_y)^2 + (n_z/L_z)^2]$

$L_x = L, \quad L_y = L_z = 2L.$ Let $\hbar^2\pi^2/8mL^2 = E_0$.

Then $E = E_0(4n_x^2 + n_y^2 + n_z^2)$; Choose the quantum numbers as follows:

n_x	n_y	n_z	E/E_0		
1	1	1	6	←	ground state
1	2	1	9	←*	* 1st 2 excited states
1	1	2	9	←*	
2	1	1	18		
1	2	2	12	←	next excited state
2	1	2	21		
2	2	1	21		
2	2	2	24		
1	1	3	14	←*	* next 2 excited states
1	3	1	14	←*	

Therefore the first 6 states are $\psi_{111}, \psi_{121}, \psi_{112}, \psi_{122}, \psi_{113}$, and ψ_{131}

with relative energies $E/E_0 = 6, 9, 9, 12, 14, 14$

---

2. $n_x = 1 \quad n_y = 1 \quad n_z = 1$

(a) $E_0 = 3\hbar^2\pi^2/2mL^2 = 3h^2/8mL^2$

$E_0 = 3(6.626 \times 10^{-34}\ \text{J·s})^2/[8(9.11 \times 10^{-31}\ \text{kg})(2 \times 10^{-10}\ \text{m})^2]$

$E_0 = 4.52 \times 10^{-18}\ \text{J} = \underline{28.2}\ \text{eV}$

(b) $E_1 = 2E_0 = \underline{56.4}\ \text{eV}$

---

3. $n^2 = 11$

(a) $E = (\hbar^2\pi^2/2mL^2)n^2 = (11/2)(\hbar^2\pi^2/mL^2)$

(b)

n_x	n_y	n_z	
2	2	1	
2	1	2	3-fold degenerate
1	2	2	

(c) $\psi_{221} = A \sin(2\pi x/L) \sin(2\pi y/L) \sin(\pi z/L)$

$\psi_{212} = A \sin(2\pi x/L) \sin(\pi y/L) \sin(2\pi z/L)$

$\psi_{122} = A \sin(\pi x/L) \sin(2\pi y/L) \sin(2\pi z/L)$

---

4. $\psi(x, y, z) = \psi_1(x)\psi_2(y)\psi_3(z)$. In the two-dimensional case,

$\psi = A(\sin k_x x)(\sin k_y y)$ where $k_x = n_x\pi/L$ and $k_y = n_y\pi/L$

$E = \hbar^2\pi^2(n_x^2 + n_y^2)/(2mL^2)$

If we let $E_0 = \hbar^2\pi^2/mL^2$, then the energy levels are :

n_x	n_y	E/E_0			
1	1	2	→	ψ_{11}	
1	2	5	→	ψ_{12}	doubly degenerate
2	1	5	→	ψ_{21}	
2	2	8	→	ψ_{22}	

---

5. (a) $n_x = n_y = n_z = 1$ and

$$E_{111} = 3h^2/8mL^2 = 3(6.6 \times 10^{-34})^2/(8 \times 1.67 \times 10^{-27} \times 2 \times 10^{-28})$$

$$= 2.45 \times 10^{-13}\text{ J} \approx \underline{1.5\text{ MeV}}$$

(b) States 211, 121, 112 have the same energy and

$$E = (2^2 + 1^2 + 1^2)h^2/8mL^2 = 2E_{111} \approx \underline{3\text{ MeV}},$$

and states 221, 122, 212 have the energy

$$E = (2^2 + 2^2 + 1^2)\ h^2/8mL^2 = 3E_{111} \approx \underline{4.5\text{ MeV}}$$

(c) Both states are threefold degenerate

---

6. $\psi(r) = [2/(4\pi)^{1/2}](1/a_0)^{3/2}e^{-r/a_0}$

(a) 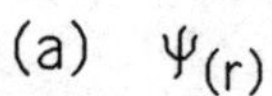

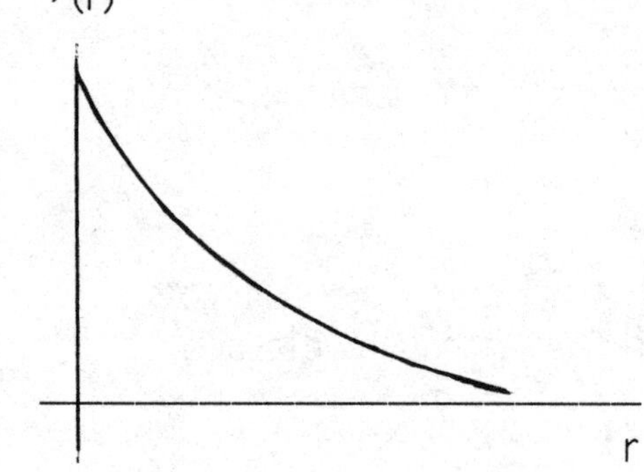

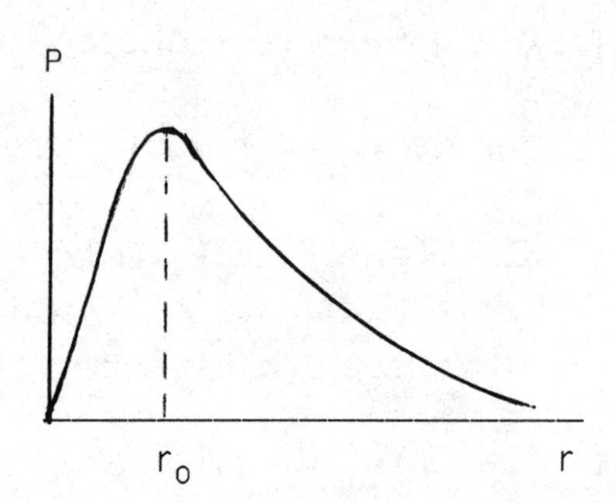

(b) The probability of finding the electron in a volume element dV is given by $|\psi|^2 dV$. Since the wave function has spherical symmetry, we see that the volume element dV is the volume of a spherical shell of radius r. That is, $dV = 4\pi r^2 dr$, so the probability of finding the electron between r and r + dr (that is, within the spherical shell) is $P = |\psi|^2 dV = 4\pi r^2 |\psi|^2 dr$.

(c) $\int |\psi|^2 dV = 4\pi \int |\psi|^2 r^2 dr$

$$= 4\pi[2/(4\pi)^{1/2}]^2(1/a_0^3) \int_0^\infty e^{-2r/a_0} r^2 dr$$

$$= (4/a_0^3) \int_0^\infty e^{-2r/a_0} r^2 dr$$

Integrating by parts, or using a Table of integrals, gives

$$\int |\psi|^2 dV = (4/a_0^3)[2/(2/a_0)^3] = 1$$

---

7. $n = 3$, Let $\ell = 0$, $m_\ell = 0$

(a) ψ_{300} corresponds to $E_{300} = - ZE_0/n^2 = - 2(13.6)/(3)^2 = - 3.02$ eV

$n = 3$; $\ell = 1$; $m_\ell = - 1, 0, 1$

ψ_{31-1} , ψ_{310} , ψ_{311} have the same average energy since n is the same.

For $n = 3$; $\ell = 2$; $m_\ell = -2 , -1, 0, 1, 2$

The states ψ_{32-2} , ψ_{32-1} , ψ_{320} , ψ_{321} , ψ_{322}

have the same energy since n is same. All states are degenerate.

---

8. $n = 1 \rightarrow \ell = 0 \rightarrow m_\ell = 0$ ψ_{100}

$\ell = 0 \rightarrow m_\ell = 0$ ψ_{200}

$n = 2 \rightarrow \ell = 1 \rightarrow m_\ell = - 1, 0, 1$ ψ_{21-1} , ψ_{210} , ψ_{211}

For $n = 2$, $E = - 13.6Z/n^2 = - 13.6(3)/4 = -$ 10.2 eV

For $n = 1$, $E = - 13.6Z = -$ 40.8 eV

---

9. (a) For positronium, $\mu = m_e/2$, so $\lambda_{32} = 656 \text{ nm} \times 2 = \underline{1312} \text{ nm}$ which is in the infrared region.

(b) For He^+, $\mu \approx m_e$, $q_1 = e$, and $q_2 = 2e$, so $\lambda_{32} = (656/4) \text{ nm} = \underline{164} \text{ nm}$ which is in the ultra-violet region.

---

10. $L = [\ell(\ell + 1)]^{1/2}\hbar$

$4.714 \times 10^{-34} = [\ell(\ell + 1)]^{1/2}(6.63 \times 10^{-34}/2\pi)$

$\ell(\ell + 1) = [(4.714 \times 10^{-34})^2(2\pi)^2]/(6.63 \times 10^{-34})^2$
$= 1.996 \times 10^1 \approx 20 = 4(4 + 1)$

so $\ell = 4$

---

11. For a d state, $\ell = 2$. Thus, m_ℓ can take on values $-2, -1, 0, 1, 2$. Since $L_z = m_\ell\hbar$, L_z can be $\pm 2\hbar$, $\pm \hbar$, and zero.

---

12. $n = 4$, $\ell = 3$, and $m_\ell = 3$.

(a) $L = [\ell(\ell + 1)]^{1/2}\hbar = [3(3 + 1)]^{1/2}\hbar = 2\sqrt{3}\,\hbar = 3.65 \times 10^{-34} \text{ J·s}$

(b) $L_z = m_\ell\hbar = 3\hbar = 3.16 \times 10^{-34} \text{ J·s}$

---

13. (a) $L = [\ell(\ell + 1)]^{1/2}\hbar$; $4.83 \times 10^{31} = [\ell(\ell + 1)]^{1/2}\hbar$, so

$\ell^2 + \ell = (4.83 \times 10^{31})^2/(1.055 \times 10^{-34})^2 \approx (4.58 \times 10^{65})^2 \approx \ell^2$

$\ell \approx 4.58 \times 10^{65}$

(b) $\Delta L/L \approx 0$ No noticeable change is detected. A quantum approach is not needed or can be detected.

---

14. (b) For the f state, $\ell = 3$.

$$L = [\ell(\ell + 1)]^{1/2}\hbar = [3(3 + 1)]^{1/2}(6.63 \times 10^{-34}/2\pi)$$

$$= 3.65 \times 10^{-34}\ \text{J}\cdot\text{s}$$

---

15. The state is 6g

(a) $n = 6$

(b) $E_n = -(13.6\ \text{eV})/n^2$ $\quad E_6 = -\ (13.6/6^2)\ \text{eV} = \underline{-\ 0.378}\ \text{eV}$

(c) A g-state $\quad \ell = 4$

$$L = [\ell(\ell + 1)]^{1/2}\hbar = (4 \times 5)^{1/2}\hbar = \underline{\sqrt{20}}\ \hbar = \underline{4.47}\ \hbar$$

(d) m_ℓ can be $-\ 4, -\ 3, -\ 2, -\ 1, 0, 1, 2, 3,$ or 4

$$L_z = m_\ell\hbar\ ; \qquad \cos\theta = L_z/L = m_\ell/[\ell(\ell + 1)]^{1/2} = m_\ell/\sqrt{2}$$

m_ℓ	-4	-3	-2	-1	0	1	2	3	4
L_z	$-4\hbar$	$-3\hbar$	$-2\hbar$	$-\hbar$	0	$\hbar$	$2\hbar$	$3\hbar$	$4\hbar$
θ	153.4°	132.1°	116.6°	102.9°	90°	77.1°	63.4°	47.9°	26.6°

---

16. When the principal quantum number is n, the following values of ℓ are possible: $\ell = 0, 1, 2, \ldots, n - 2, n - 1$. For a given value of ℓ, there are $2\ell + 1$ possible values of m_ℓ. For each value of m_ℓ there are 2 possible values of m_s; therefore two possible electrons. The maximum number of electrons which can be accommodated in the nth

level is therefore:

$$2 \times [(2(0) + 1) + 2(1) + 1) + \ldots + (2\ell + 1) + \ldots + (2(n-1) + 1)]$$

$$= 2 \times [2 \sum_{\ell=0}^{n-1} \ell + \sum_{\ell=0}^{n-1} 1] = (2 \sum_{\ell=0}^{n-1} \ell + n)(2)$$

But $\sum_{\ell=0}^{k} \ell = k(k+1)/2$ so the maximum number of electrons to be accommodated is $2[2(n-1)n/2 + n] = 2n^2$.

---

17. $\psi_{1s}(r) = [1/(\pi a_0^3)^{1/2}]e^{-r/a_0}$

$$\partial\psi_{1s}/\partial x = -(1/a_0)[1/(\pi a_0^3)^{1/2}]e^{-r/a_0}(\partial r/\partial x)$$

$$= -(1/a_0)(x/r)[1/(\pi a_0^3)^{1/2}]e^{-r/a_0}$$

In the above expression we used $r = (x^2 + y^2 + z^2)^{1/2}$ and

$(\partial r/\partial x) = x/(x^2 + y^2 + z^2)^{1/2} = x/r$

Therefore $(\partial^2\psi/\partial x^2) = (-1/a_0 r)[1/(\pi a_0^2)^{1/2}]e^{-r/a_0}$

$$+ (1/a_0)(x^2/r^3)[1/(\pi a_0^3)^{1/2}]e^{-r/a_0} + (x^2/a_0^2 r^2)[1/(\pi a_0^3)^{1/2}]e^{-r/a_0}$$

$$= [(-1/a_0 r) + x^2(1/a_0 r^3 + 1/a_0^2 r^2)]\psi_{1s}$$

Then $(\partial^2\psi_{1s}/\partial x^2) + (\partial^2\psi_{1s}/\partial y^2) + (\partial^2\psi_{1s}/\partial z^2)$

$$= [(-3/a_0 r) + (x^2 + y^2 + z^2)(1/a_0 r^3 + 1/a_0^2 r^2)]\psi_{1s}$$

$$= [(-3/a_0 r) + 1/a_0 r + 1/a_0^2]\psi_{1s} = [(-2/a_0 r) + 1/a_0^2]\psi_{1s}$$

But $(\partial^2\psi/\partial x^2) + (\partial^2\psi/\partial y^2) + (\partial^2\psi/\partial z^2) = (2m/\hbar^2)(-ke^2/r - E)\psi_{1s}$

so $(-2/a_0 r) + 1/a_0^2 = (2m/\hbar^2)(-ke^2/r - E)$

Then $1/a_0 = mke^2/\hbar^2$ and $E = -\hbar^2/2ma_0^2 = (-\hbar^2/2m)(mke^2/\hbar^2)^2$

or $E = -\, mk^2e^4/2\hbar^2$

---

18. (a) $\psi_{2s}(r) = [1/4(2\pi)^{1/2}](1/a_0)^{3/2}\,(2 - r/a_0)e^{-r/2a_0}$

At $r = a_0 = 0.529 \times 10^{-10}$ m, we find

$$\psi_{2s}(a_0) = [1/4(2\pi)^{1/2}](1/a_0)^{3/2}\,(2 - 1)e^{-1/2} = (0.380)(1/a_0)^{3/2}$$

$$= (0.380)[1/(0.529 \times 10^{-10}\text{ m})]^{3/2} = \underline{9.88 \times 10^{14}}\text{ m}^{-3/2}$$

(b) $|\psi_{2s}(a_0)|^2 = (9.88 \times 10^{14}\text{ m}^{-3/2})^2 = \underline{9.75 \times 10^{29}}\text{ m}^{-3}$

(c) Using the result to part (b), we get

$$P_{2s}(a_0) = 4\pi a_0^2|\psi_{2s}(a_0)|^2 = \underline{3.43 \times 10^{10}}\text{ m}^{-1}$$

---

19. $R_{2p}(r) = Are^{-r/2a_0}$ where $A = 1/\{2(6)^{1/2}\,a_0^{5/2}\}$

$P(r) = r^2R_{2p}^2(r) = A^2r^4e^{-r/a_0}$

$$\langle r\rangle = \int_0^\infty rP(r)dr = A^2\int_0^\infty r^5e^{-r/a_0} = A^2a_0^6 5! = \underline{5a_0}$$

---

20. (a) $1/\alpha = \hbar c/ke^2 = (6.63 \times 10^{-34})(3 \times 10^8)/2\pi(9 \times 10^9)(1.6 \times 10^{-19})^2$

$1/\alpha = \underline{137.036}$

(b) $\lambda_0/r_e = (h/mc)\,(e^2/mc^2) = he^2/c = 2\pi/\alpha = 2\pi \times 137$

(c) $a_0/\lambda_C = (\hbar^2/mke^2)(\hbar/mc) = (1/2\pi)\hbar c/ke^2 = 137/2\pi = 1/2\pi\alpha$

(d) $1/Rr_0 = (mke^2/\hbar^2)(4\pi c\hbar^3/mk^2e^4) = 4\pi\hbar c/ke^2 = 4\pi(137) = 4\pi/\alpha$

Chapter 7

21. There is no force on a free particle, so that $U(r)$ is a constant which, for simplicity, we take to be zero. Substituting

$\Psi(r,t) = \psi_1(x)\psi_2(y)\psi_3(z)\phi(t)$ into Schrodinger's equation with $U(r) = 0$

$$-(\hbar^2/2m)\{\partial^2/\partial x^2 + \partial^2/\partial y^2 + \partial^2/\partial z^2\}\Psi(r,t) = i\hbar(\partial/\partial t)\Psi(r,t)$$

and dividing through by $\psi_1(x)\psi_2(y)\psi_3(z)\phi(t)$ gives

$$-(\hbar^2/2m)\{\psi_1''(x)/\psi_1(x) + \psi_2''(y)/\psi_2(y) + \psi_3''(z)/\psi_3(z)\} = i\hbar\,\phi'(t)/\phi(t)$$

Each term in this equation is a function of one variable only. Since the variables x,y,z,t are all independent, each term, by itself, must be a constant, an observation which leads to the four separate equations

$$-(\hbar^2/2m)\{\psi_1''(x)/\psi_1(x)\} = E_1$$

$$-(\hbar^2/2m)\{\psi_2''(y)/\psi_2(y)\} = E_2$$

$$-(\hbar^2/2m)\{\psi_3''(z)/\psi_3(z)\} = E_3$$

$$i\hbar\{\phi'(t)/\phi(t)\} = E$$

subject to the condition that

$$E_1 + E_2 + E_3 = E$$

The equation for ψ_1 can be rearranged as

$$d^2\psi_1/dx^2 = (-2mE_1/\hbar^2)\psi_1(x)$$

whereupon it is evident the solutions are sinusoidal

$$\psi_1(x) = \alpha_1 \sin(k_1x) + \beta_1 \cos(k_1x)$$

with $k_1{}^2 = 2mE_1/\hbar^2$. However, the <u>mixing coefficients α_1 and β_1</u> are indeterminate from this analysis. Similarly, we find

$$\psi_2(y) = \alpha_2 \sin(k_2 y) + \beta_2 \cos(k_2 y)$$

$$\psi_3(z) = \alpha_3 \sin(k_3 z) + \beta_3 \cos(k_3 z)$$

with $k_2{}^2 = 2mE_2/\hbar^2$ and $k_3{}^2 = 2mE_3/\hbar^2$. The equation for ϕ can be integrated once to get

$$\phi(t) = \gamma\, e^{-i\omega t}$$

with $\omega = E/\hbar$ and γ another undetermined coefficient.

Since the energy operator is $[E] = i\hbar\partial/\partial t$ and

$$i\hbar(\partial/\partial t)\phi = E\phi$$

energy is sharp at the value E in this state. Also, since

$[p_x{}^2] = -\hbar^2(\partial^2/\partial x^2)$ and

$$-\hbar^2(\partial^2/\partial x^2)\psi_1 = (\hbar k_1)^2\psi_1$$

the magnitude of momentum in the x direction is sharp at the value $\hbar k_1$. Similarly, the magnitude of momentum in the y and z directions are sharp at the values $\hbar k_2$ and $\hbar k_3$, respectively. This differs from the free particle waves of Example 7.3 where both the magnitude and sign of momentum were sharp in each of the three directions x, y, and z. The sign of momentum will be sharp here only if the mixing coefficients are chosen in the ratios $\alpha_1/\beta_1 = i$, etc.

---

22. The stationary states for a particle in a cubic box are, from Eq. 7.15

$$\Psi(r,t) = A \sin(k_1x)\sin(k_2y)\sin(k_3z)e^{-iEt/\hbar} \qquad 0 \leq x, y, z \leq L$$

$$= 0 \text{ elsewhere}$$

where $k_1 = n_1\pi/L$, etc. Since Ψ is nonzero only for $0 < x < L$, etc., the normalization condition Eq. (7.2) reduces to an integral over the volume of a cube with one corner at the origin:

$$1 = \int dx \int dy \int dz \left|\Psi(r,t)\right|^2$$

$$= A^2\left\{\int_0^L \sin^2(k_1x)dx \int_0^L \sin^2(k_2y)dy \int_0^L \sin^2(k_3z)dz\right\}$$

Using $2\sin^2\theta = 1 - \cos2\theta$ gives

$$\int_0^L \sin^2(k_1x)dx = L/2 - (1/4k_1)\sin(2k_1x)\Big|_0^L$$

But $k_1L = n_1\pi$, so the last integral on the right is zero. The same result is obtained for the integrations over y and z. Thus, normalization requires

$$1 = A^2(L/2)^3$$

or $A = (2/L)^{3/2}$ for any of the stationary states.

Allowing the edge lengths to be different at L_1, L_2, and L_3 requires only that L^3 be replaced by the box volume $L_1L_2L_3$ in the final result:

$$A = \left\{(2/L_1)(2/L_2)(2/L_3)\right\}^{1/2} = \{8/L_1L_2L_3\}^{1/2}$$

This follows because it is still true that the wave must vanish at the walls of the box, so that $k_1L_1 = n_1\pi$, etc.

24. Inside the box the electron is free, and so has momentum and energy given by the de Broglie relations

$$|\mathbf{p}| = \hbar|\mathbf{k}| \qquad \text{and} \qquad E = \hbar\omega$$

with

$$E = (c^2|\mathbf{p}|^2 + m^2c^4)^{1/2}$$

for this, the relativistic case. Here $k = (k_1, k_2, k_3)$ is the wavevector whose components k_1, k_2, and k_3 are wavenumbers along each of three mutually perpendicular axes. In order for the wave to vanish at the walls, the box must contain an integral number of half-wavelengths in each direction. Since $\lambda_1 = 2\pi/k_1$ etc., this gives

$$L = n_1(\lambda_1/2) \qquad \text{or} \qquad k_1 = n_1\pi/L$$

$$L = n_2(\lambda_2/2) \qquad \text{or} \qquad k_2 = n_2\pi/L$$

$$L = n_3(\lambda_3/2) \qquad \text{or} \qquad k_3 = n_3\pi/L$$

Thus,

$$|\mathbf{p}|^2 = \hbar^2|\mathbf{k}|^2 = \hbar^2\{k_1^2 + k_2^2 + k_3^2\} = (\pi\hbar/L)^2\{n_1^2 + n_2^2 + n_3^2\}$$

and the allowed energies are

$$E = [(\pi\hbar c/L)^2\{n_1^2 + n_2^2 + n_3^2\} + (mc^2)^2]^{1/2}$$

For the ground state $n_1 = n_2 = n_3 = 1$. For an electron confined to

$L = 10$ fm, we use $m = 0.511\ \text{MeV}/c^2$ and $\hbar c = 197.3$ MeV·fm to get

$$E = \left\{3[(\pi)(197.3\ \text{MeV·fm})/(10\ \text{fm})]^2 + (0.511\ \text{MeV})^2\right\}^{1/2} = \underline{107.4}\ \text{MeV}$$

---

25. To find Δr we first compute $\langle r^2 \rangle$ using the radial probability density for the 1s state of hydrogen:

$$P_{1s}(r) = (4/a_0^3)r^2 e^{-2r/a_0}$$

Then

$$\langle r^2 \rangle = \int_0^\infty r^2 P_{1s}(r)dr = (4/a_0^3)\int_0^\infty r^4 e^{-2r/a_0}\,dr$$

With $z = 2r/a_0$, this is

$$\langle r^2 \rangle = (4/a_0^3)(a_0/2)^5 \int_0^\infty z^4 e^{-z}dz$$

The integral on the right is (see Example 7.13)

$$\int_0^\infty z^4 e^{-z}dz = 4!$$

so that

$$\langle r^2 \rangle = (4/a_0^3)(a_0/2)^5(4!) = 3a_0^2$$

and

$$\Delta r = (\langle r^2 \rangle - \langle r \rangle^2)^{1/2} = [3a_0^2 - (1.5a_0)^2]^{1/2} = 0.866\,a_0$$

Since Δr is an appreciable fraction of the average distance, the whereabouts of the electron are largely unknown in this case.

---

26. The most probable distance is the value of r which maximizes the radial probability $P(r) = |rR(r)|^2$. Since P(r) is largest where rR(r) reaches its maximum, we look for the most probable distance by setting $d\{rR(r)\}/dr$ equal to zero, using the functions R(r) from Table 7.4. For clarity, we measure distances in bohrs, so that r/a_0 becomes simply r, etc. Then for the 2s state of hydrogen, the condition for a maximum is

$$0 = (d/dr)\{(2r - r^2)e^{-r/2}\} = \{2 - 2r - (1/2)(2r - r^2)\}e^{-r/2}$$

or

$$0 = 4 - 6r + r^2$$

There are two solutions, which may be found by completing the square to get

$$0 = (r - 3)^2 - 5 \quad \text{or} \quad r = 3 \pm \sqrt{5} \text{ bohrs}$$

Of these, $r = 3 + \sqrt{5} = 5.236\,a_0$ gives the largest value of P(r), and so is the most probable distance.

For the 2p state of hydrogen, a similar analysis gives

$$0 = (d/dr)\{r^2e^{-r/2}\} = \{2r - (1/2)r^2\}e^{-r/2}$$

with the obvious roots r = 0 (a minimum) and r = 4 (a maximum). Thus, the most probable distance for the 2p state is $r = 4a_0$, in agreement with the simple Bohr model.

---

27. The probabilities are found by integrating the radial probability density for each state. P(r), from r = 0 to $r = 4a_0$. For the 2s state we find from Table 7.4 (with Z = 1 for hydrogen)

$$P_{2s}(r) = |rR_{2s}(r)|^2 = (8a_0)^{-1}(r/a_0)^2(2 - r/a_0)^2e^{-r/a_0}$$

and

$$P = (8a_o)^{-1} \int_o^{4a_0} (r/a_o)^2(2 - r/a_o)^2 e^{-r/a_0}\, dr$$

Changing variables from r to $z = r/a_o$ gives

$$P = 8^{-1} \int_o^4 (4z^2 - 4z^3 + z^4)e^{-z}\, dz$$

Repeated integration by parts gives

$$P = 8^{-1} \left\{- (4z^2 - 4z^3 + z^4) - (8z - 12z^2 + 4z^3) - (8 - 24z + 12z^2) - (-24 + 24z) - (24)\right\}e^{-z} \Big|_o^4$$

$$= 8^{-1} \left\{- (64 + 96 + 104 + 72 + 24)e^{-4} + 8\right\} = \underline{0.176}$$

For the 2p state of hydrogen

$$P_{2p}(r) = \left|r R_{2p}(r)\right|^2 = (24a_o)^{-1} (r/a/_o)^4 e^{-r/a_0}$$

and

$$P = (24a_o)^{-1} \int_o^{4a_0} (r/a_o)^4 e^{-r/a_0}\, dr = 24^{-1} \int_o^4 z^4 e^{-z}\, dz$$

Again integrating by parts, we get

$$P = 24^{-1}\left\{-z^4 - 4z^3 - 12z^2 - 24z - 24\right\}e^{-z}\Big|_o^4 = 24^{-1}\left\{-824e^{-4} + 24\right\} = \underline{0.371}$$

The probability for the 2s electron is much smaller, suggesting that this electron spends more of its time in the outer regions of the atom. This is in accord with classical physics, where the electron in a lower angular momentum state is described by orbits more elliptic in shape.

Chapter 7

28. The averages $\langle r \rangle$ and $\langle r^2 \rangle$ are found by weighting the probability density for this state

$$P_{1s}(r) = 4(Z/a_0)^3 r^2 e^{-2Zr/a_0}$$

with r and r^2, respectively, in the integral from r = 0 to r = ∞:

$$\langle r \rangle = \int_0^\infty r P_{1s}(r) dr = 4(Z/a_0)^3 \int_0^\infty r^3 e^{-2Zr/a_0} dr$$

$$\langle r^2 \rangle = \int_0^\infty r^2 P_{1s}(r) dr = 4(Z/a_0)^3 \int_0^\infty r^4 e^{-2Zr/a_0} dr$$

Substituting $z = 2Zr/a_0$ gives

$$\langle r \rangle = 4(Z/a_0)^3 (a_0/2Z)^4 \int_0^\infty z^3 e^{-z} dz = (3!/4)(a_0/Z) = (3/2)a_0/Z$$

$$\langle r^2 \rangle = 4(Z/a_0)^3 (a_0/2Z)^5 \int_0^\infty z^4 e^{-z} dz = (4!/8)(a_0/Z)^2 = 3(a_0/Z)^2$$

and

$$\Delta r = (\langle r^2 \rangle - \langle r \rangle^2)^{1/2} = (a_0/Z)[3 - 9/4]^{1/2} = 0.866(a_0/Z)$$

The momentum uncertainty is deduced from the average potential energy

$$\langle U \rangle = -kZe^2 \int_0^\infty (1/r) P_{1s}(r) dr = -4kZe^2 (Z/a_0)^3 \int_0^\infty r e^{-2Zr/ao} dr$$

$$= -4kZe^2 (Z/a_0)^3 (a_0/2Z)^2 = -k(Ze)^2/a_0$$

Then, since

$$E = -k(Ze)^2/2a_0$$

for the 1s level, and $a_0 = \hbar^2/mke^2$, we obtain

$$\langle p^2\rangle = 2m\langle K\rangle = 2m(E - \langle U\rangle) = 2mk(Ze)^2/2a_0 = (Z\hbar/a_0)^2$$

With $\langle p\rangle = 0$ from symmetry, we get

$$\Delta p = (\langle p^2\rangle)^{1/2} = Z\hbar/a_0$$

and

$$\Delta r\Delta p = 0.866\hbar$$

for any Z, consistent with the uncertainty principle.

---

29. Outside the surface, $U(x) = -A/x$ (to give $F = -dU/dx = -A/x^2$), and Schrodinger's equation is

$$-(\hbar^2/2m)\, d^2\psi/dx^2 + (-A/x)\psi(x) = E\psi(x)$$

From Equation 7.19 $g(r) = rR(r)$ satisfies a one-dimensional Schrodinger equation with effective potential

$$U_{eff}(r) = U(r) + \ell(\ell + 1)\hbar^2/2mr^2$$

With $\ell = 0$ (s states) and $U(r) = -kZe^2/r$, the equation for $g(r)$ has the same form as that for $\psi(x)$. Furthermore, $\psi(0) = 0$ if no electrons can cross the surface, while $g(0) = 0$ since $R(0)$ must be finite. It follows that the functions $g(r)$ and $\psi(x)$ are the same, and that the energies in the present case are the hydrogenic levels

$$E_n = -(Z^2ke^2/2a_0)[1/n^2]$$

with the replacement $kZe^2 \rightarrow A$. Remembering that $a_0 = \hbar^2/mke^2$, we get

$$E_n = -(mA^2/2\hbar^2)[1/n^2] \qquad n = 1, 2, \ldots$$

Chapter 8

1. $n = 2$; $\ell = 1, 0$; $s = 1/2$

   $j = 1 + 1/2$, $1 - 1/2$, $0 + 1/2$, $|0 - 1/2|$

   Let $j = 1.5$; $m_j = -1.5, -0.5, 0.5, 1.5$

   $j = 0.5$; $m_j = -0.5, 0.5$

---

2. For a d electron, $\ell = 2$; $s = 1/2$

   $j = 2 + 1/2$, $2 - 1/2$

   For $j = 2.5$

   $m_j = -2.5, -1.5, -0.5, 0.5, 1.5, 2.5$

   For $j = 1.5$

   $m_j = -1.5, -0.5, 0.5, 1.5$

---

3. (a) $\ell = 4$ is a g state, so $7G_{9/2}$

   (b) $\ell = 5$ is an h state

   If $\ell = 5$, j can be $\ell + s = 5 + 1/2 = 11/2$ or $\ell - s = 5 - 1/2 = 9/2$

   Therefore we have $6H_{9/2}$ and $6H_{11/2}$

---

4. (a) $4F_{5/2} \rightarrow n = 4$, $\ell = 3$, $j = 5/2$

   (b) $J = [j(j + 1)]^{1/2}\hbar = [(5/2)(7/2)]^{1/2}\hbar = [(35)/4]^{1/2}\hbar = \underline{2.96}\hbar$

   (c) $J_z = m_j\hbar$ where m_j can be $-j, -j + 1, \ldots, j - 1, j$

   so here m_j can be $-5/2, -3/2, -1/2, 1/2, 3/2, 5/2$

   J_z can be $-(5/2)\hbar$, $-(3/2)\hbar$, $-(1/2)\hbar$, $(1/2)\hbar$, $(3/2)\hbar$, or $(5/2)\hbar$

5. (a) $n = 1$; For $n = 1$, $\ell = 0$, $m_\ell = 0$, $m_s = \pm 1/2$ → 2 sets

n	ℓ	m_ℓ	m_s
1	0	0	-1/2
1	0	0	+1/2

$2n^2 = 1(1)^2 = 2$

(b) For $n = 2$ we have

n	ℓ	m_ℓ	m_s
2	0	0	±1/2
2	1	-1	±1/2
2	1	0	±1/2
2	1	1	±1/2

yields 8 sets ; $2n^2 = 2(2)^2 = 8$

Note that the number is twice the number of m_ℓ values . Also that for each ℓ there are $2\ell + 1$ m_ℓ values . Finally , ℓ can take on values ranging from 0 to $n - 1$. So the general expression is

$$s = \sum_{0}^{n-1} 2(2\ell + 1)$$

The series is an arithmetic progression: $2 + 6 + 10 + 14 \ldots$, the sum of which is

$$s = (n/2)[2a + (n^{-1})d] \quad \text{where} \quad a = 2, \quad d = 4.$$

$$s = (n/2)[4 + (n^{-1})4] = 2n^2$$

(c) $n = 3$: $2(1) + 2(3) + 2(5) = 2 + 6 + 10 = 18 = 2n^2 = 2(3)^2 = 18$

(d) $n = 4$: $2(1) + 2(3) + 2(5) + 2(7) = 32 = 2n^2 = 2(4)^2 = 32$

(e) $n = 5$: $32 + 2(9) = 32 + 18 = 50 = 2n^2 = 2(5)^2 = 50$

6. (a) $1s^2 2s^2 2p^4$

(b) For the two 1s electrons, $n = 1$, $\ell = 0$, $m_\ell = 0$, $m_s = \pm 1/2$

For the two 2s electrons, $n = 2$, $\ell = 0$, $m_\ell = 0$, $m_s = \pm 1/2$

For the four 2p electrons, $n = 2$, $\ell = 1$, $m_\ell = 1, 0, -1$, $m_s = \pm 1/2$

---

7. $[Ar]3d^4 4s^2$ is more energetic since one electron has a higher n number than $[Ar]3d^5 4s$. $[Ar]3d^5 4s$ has 2 more unpaired electron spins; the element is Cr($[Ar]3d^5 4s$). This supports Hund's rule since $[Ar]3d^5 4s$ has 6 unpaired spins compared to 4 for $[Ar]3d^4 4s^2$.

---

8. (a) $[Kr]\ 4d^9 5s$ is more energetic than $[Kr]\ 4d^{10}$ since an electron in the first case has a higher n value. (Recall $E \propto 1/n^2$, so that a higher n means a smaller $1/n^2$ but a larger $1/n^2$.)

(b) All spins are paired for $[Kr]\ 4d^{10}$ and two are unpaired for $[Kr]\ 4d^9 5s$

(c) $[Kr]\ 4d^{10}$ is preferred even though it violates Hund's rule since it has less energy and causes a sublevel to be filled.

(d) The element is Pd.

---

10. (b) 3p subshell. For a p state, $\ell = 1$. Thus m_ℓ can take on values $-\ell$ to ℓ, or $-1, 0, 1$. For each m_ℓ, m_s can be $\pm 1/2$.

n	ℓ	m_ℓ	m_s
3	1	-1	-1/2
3	1	-1	+1/2
3	1	0	-1/2
3	1	0	+1/2
3	1	1	-1/2
3	1	1	+1/2

---

11. (a) A d-subshell has $\ell = 2$. Then L_z = can be $(-2, -1, 0, 1, \text{ or } 2)\hbar$

J_z can be $(-5/2, -3/2, -1/2, 1/2, 3/2, \text{ or } 5/2)\hbar$ for each electron.

(b) For the entire atom $J_z(\text{max}) = J_{z1} + J_{z2} = 5/2\ \hbar + 3/2\ \hbar = 4\hbar$ so possible J_z values are $\pm(0, 1, 2, 3, \text{ or } 4)\hbar$. Thus the possible J_z values are $(-4, -3, -2, -1, 0, 1, 2, 3, \text{ and } 4)\hbar$ and the largest total angular momentum quantum number is 4.

(c) In this case the magnitude of the total angular momentum is

$$|J| = [J(J+1)]^{1/2}\hbar = (20)^{1/2}\hbar = \underline{4.72 \times 10^{-34}}\ \text{J}\cdot\text{s}$$

---

12. $\Delta E = 2\mu_B B = hf$

$$2(9.27 \times 10^{-24})(0.715) = (6.63 \times 10^{-34})f$$

so $f = \underline{2.00 \times 10^{10}}$ Hz

---

14. (a) $\mathbf{J}\cdot\mathbf{J} = (\mathbf{L} + \mathbf{S})\cdot(\mathbf{L} + \mathbf{S}) \quad \Rightarrow \quad J^2 = L^2 + S^2 + 2\mathbf{L}\cdot\mathbf{S}$

$$\mathbf{L}\cdot\mathbf{S} = (1/2)(J^2 - L^2 - S^2) = (\hbar^2/2)[j(j+1) - \ell(\ell+1) - s(s+1)]$$

(b) $\mathbf{L}\cdot\mathbf{S} = |\mathbf{L}|\,|\mathbf{S}|\cos\theta = \hbar^2[\ell(\ell+1)]^{1/2}[s(s+1)]^{1/2}\cos\theta$ so

$$\cos\theta = (1/2)[j(j+1) - \ell(\ell+1) - s(s+1)]/[\ell(\ell+1)]^{1/2}[s(s+1)]^{1/2}$$

since $s = 1/2$,

$$\cos\theta = [j(j+1) - \ell(\ell+1) - 3/4]/(3)^{1/2}[\ell(\ell+1)]^{1/2}$$

i) $P_{1/2}$ has $\ell = 1$, $j = 1/2$

$$\cos\theta = [(1/2)(3/2) - 1(1+1) - 3/4]/\sqrt{3}\sqrt{2} = -(2/3)^{1/2} = -0.816,$$

$\theta = \underline{144.7°}$

$P_{3/2}$ has $\ell = 1$, $j = 3/2$

$\cos\theta = [(3/2)(5/2) - 1(1 + 1) - 3/4]/\sqrt{3}\sqrt{2} = 1/\sqrt{6} = 0.408$

$\theta = \underline{65.9°}$

ii) $H_{9/2}$ has $\ell = 5$, $j = 9/2$

$\cos\theta = [(9/2)(11/2) - 5(6) - 3/4]/[5(6)]^{1/2}\sqrt{3} = -6/\sqrt{90} = -2/\sqrt{10}$

$\cos\theta = -0.632$, $\theta = \underline{129.2°}$

$H_{11/2}$ has $\ell = 5$, $j = 11/2$

$\cos\theta = [(11/2)(13/2) - 5(6) - 3/4]/[5(6)]^{1/2}\sqrt{3} = 5/\sqrt{90} = 0.527$

$\theta = \underline{58.2°}$

---

15. From Equation 7.8 we have $E = (\hbar^2\pi^2/2mL^2)(n_x^2 + n_y^2 + n_z^2)$

$E = (1.054 \times 10^{-34})^2(\pi^2)(n_x^2 + n_y^2 + n_z^2)/[2\,(9.11 \times 10^{-31})(2 \times 10^{-10})^2]$

$= (1.5 \times 10^{-18}\text{ J})(n_x^2 + n_y^2 + n_z^2) = (9.4\text{ eV})(n_x^2 + n_y^2 + n_z^2)$

(a) 2 electrons per state. The lowest states have

$(n_x, n_y, n_z) = (1, 1, 1) \Rightarrow E_{111} = (9.4\text{ eV})(1^2 + 1^2 + 1^2)\text{eV} = 28.2\text{ eV}$

For $(n_x, n_y, n_z) = (1, 1, 2)$ or $(1, 2, 1)$ or $(2, 1, 1)$,

$E_{112} = E_{121} = E_{211} = (9.4\text{ eV})(1^2 + 1^2 + 2^2) = 56.4\text{ eV}$

$E_{min} = 2 \times (E_{111} + E_{112} + E_{121} + E_{211}) = 2(28.2 + 3 \times 56.5) = \underline{394.8}\text{ eV}$

(b) All 8 particles go into the $(n_x, n_y, n_z) = (1, 1, 1)$ state, so

$E_{min} = 8 \times E_{111} = \underline{225.6}\text{ eV}$

---

16. Since $\mathbf{L} = \mathbf{r} \times \mathbf{p}$, and $\mathbf{p} = m d\mathbf{r}/dt$ is tangent to the path at every point, the orbit is confined to the plane perpendicular to $\mathbf{L}$. The Figure below shows the positions of the particle in this orbital plane at consecutive times t and t + dt. The area (of the triangle) swept out in this time is

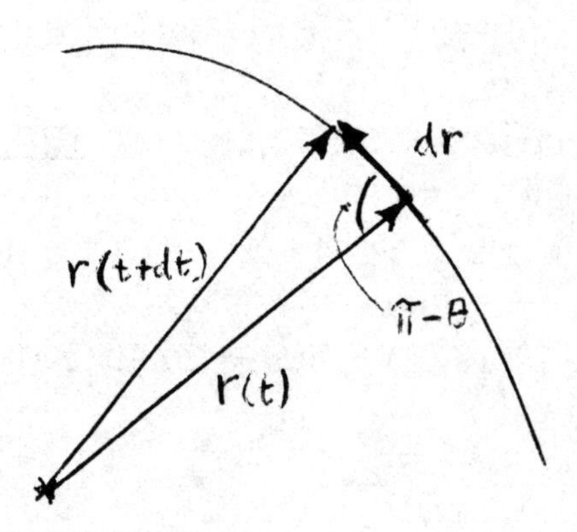

$$dA = (1/2)|\mathbf{r}||d\mathbf{r}|\sin(\pi - \theta) = (1/2)|\mathbf{r}||d\mathbf{r}|\sin(\theta) = (1/2)|\mathbf{r} \times d\mathbf{r}|$$

Then

$$dA/dt = (1/2)|\mathbf{r} \times d\mathbf{r}/dt| = (1/2m)|\mathbf{r} \times \mathbf{p}| = (1/2m)|\mathbf{L}|$$

---

17. The exiting beams differ in the spin orientation of the outermost atomic electron. The energy difference derives from the magnetic energy of this spin in the applied field $\mathbf{B}$:

$$U = -\boldsymbol{\mu}_s \cdot \mathbf{B} = g(-e/2m)S_z B = -g\mu_B B m_s$$

With g = 2 for electrons, the energy difference between the up spin ($m_s = 1/2$) and down spin ($m_s = -1/2$) orientations is

$$\Delta U = g\mu_B B = (2)(9.273 \times 10^{-24}\ \text{J/T})(0.5\ \text{T}) = 9.273 \times 10^{-24}\ \text{J}$$
$$= \underline{5.80 \times 10^{-5}}\ \text{eV}.$$

Chapter 8

18. The angular momentum **L** of a spinning ball is related to the angular velocity of rotation $\boldsymbol{\omega}$ as

$$L = I\boldsymbol{\omega}$$

where I, the moment of inertia, is given in terms of the mass m and radius R of the ball as

$$I = (2/5)mR^2$$

For the electron this gives

$$I = (2/5)(511 \times 10^3 \text{ eV}/c^2)(3 \times 10^{-6} \text{ nm})^2 = 1.840 \times 10^{-6} \text{ eV}\cdot\text{nm}^2/c^2$$

Then, using $L = (\sqrt{3}/2)\hbar$, we find

$$\omega = L/I = (\sqrt{3}/2)(197.3 \text{ eV}\cdot\text{nm}/c)/(1.840 \times 10^{-6} \text{ eV}\cdot\text{nm}^2/c^2)$$
$$= 9.286 \times 10^7 \text{ c/nm}.$$

The equatorial speed is

$$v = R\omega = (3 \times 10^{-6} \text{ nm})(9.286 \times 10^7 \text{ c/nm}) = 278.6c.$$

---

19. We find L from $\mathbf{L} = I\boldsymbol{\omega}$, where I is the moment of inertia for the cylinder about its longitudinal axis. To find I, consider first a cylindrical shell with radius r, thickness dr, and the same length as the cylinder. The shell has mass proportional to its volume:

$$dM/M = dV/V = 2\pi r dr/\pi R^2 \quad \text{or} \quad dM = M(2r/R^2)dr$$

Since all the shell mass is at the same perpendicular distance r from the rotation axis, the shell moment of inertia is

$$dI = r^2 dM = M(2r^3/R^2)dr$$

Integrating r from zero to the cylinder radius R gives the moment of inertia for the solid cylinder:

$$I = \int dI = (2M/R^2)\int_0^R r^3 dr = (1/2)MR^2$$

Then

$$\mathbf{L} = I\boldsymbol{\omega} = (1/2)MR^2\boldsymbol{\omega}$$

The direction of **L** is that of ω, or along the axis of rotation with a sense given by the right hand rule.

The charge Q on the curved surface of the rotating cylinder simulates a circular sheet of current. All of this charge passes a fixed line in the time it takes the cylinder to rotate once, $T = 2\pi/\omega$, so the effective current is

$$i = Q/T = Q\omega/2\pi$$

and the magnetic moment of the rotating cylinder becomes

$$\mu = iA = (Q\omega/2\pi)\pi R^2 = QR^2\omega/2$$

The direction of $\boldsymbol{\mu}$ is that of $\boldsymbol{\omega}$, so that

$$\boldsymbol{\mu} = (QR^2/2)\boldsymbol{\omega}$$

Comparing the results for $\boldsymbol{\mu}$ and **L**, we see they are related as

$$\boldsymbol{\mu} = (Q/M)\,\mathbf{L}$$

implying a g-factor of 2 for this object.

---

20. With s = 3/2, the spin magnitude is

$$S = [s(s + 1)]^{1/2}\hbar = ([15]^{1/2}/2)\hbar$$

The z-component of spin is $S_z = m_s\hbar$ where m_s ranges from -s to s in integer steps or, in this case,

$$m_s = -3/2, -1/2, +1/2, +3/2$$

The spin vector S is inclined to the z-axis by an angle θ such that

$$\cos(\theta) = S_z/|S| = m_s\hbar/([15]^{1/2}/2)\hbar = m_s/([15]^{1/2}/2)$$

$$= -3/(15)^{1/2}, \ -1/(15)^{1/2}, \ +1/(15)^{1/2}, \ +3/(15)^{1/2}$$

or

$$\theta = 140.77°, 104.96°, 75.04°, 39.23°$$

The Ω^- does obey the Pauli Exclusion Principle, since the spin s of this particle is half-integral, as it is for all fermions.

---

21. The spin of the atomic electron has a magnetic energy in the field of the orbital moment given by Equations 8.7 and 8.10 with a g-factor of 2, or

$$U = -\mu_s \cdot B = 2(e/2m)S_z B = 2\mu_B m_s B$$

The magnetic field **B** originates with the orbiting electron. To estimate **B**, we adopt the equivalent viewpoint of the atomic nucleus (proton) circling the electron, and borrow a result from classical electromagnetism for the **B** field at the center of a circular current loop with radius r:

$$B = 2k_m\mu/r^3$$

Here k_m is the magnetic constant and $\mu = i\pi r^2$ is the magnetic moment of the loop, assuming it carries a current i. In the atomic case, we identify r with the orbit radius and the current i with the proton charge +e divided by the orbital period $T = 2\pi r/v$. Then

$$\mu = evr/2 = (e/2m)L$$

where $L = mvr$ is the orbital angular momentum of the electron. For a p electron $\ell = 1$ and $L = [\ell(\ell + 1)]^{1/2}\hbar = \sqrt{2}\hbar$, so

$$\mu = (e\hbar/2m)\sqrt{2} = \mu_B\sqrt{2} = 1.31 \times 10^{-23} \text{ J/T}$$

For r we take a typical atomic dimension, say $4a_0$ ($= 2.12 \times 10^{-10}$ m) for a 2p electron, and find

$$B = 2(10^{-7})(1.31 \times 10^{-23})/(2.12 \times 10^{-10})^3 = 0.276 \text{ T}$$

Since m_s is $\pm 1/2$, the magnetic energy of the electron spin in this field is

$$U = \pm\ \mu_B B = \pm\ (9.27 \times 10^{-24} \text{ J/T})(0.276 \text{ T}) = \pm\ 2.56 \times 10^{-24} \text{ J}$$

$$= \pm\ 1.59 \times 10^{-5} \text{ eV}$$

The up spin orientation (+) has the higher energy; the predicted energy difference between the up (+) and down (-) spin orientations is twice this figure, or about 3.18×10^{-5} eV, a result which compares favorably with the measured value, 5×10^{-5} eV.

---

22. Classically, $|\mathbf{L}| = |\mathbf{r}|p_\perp$, where $p_\perp$ is the component of particle momentum perpendicular to **r**. Remembering that **p** is tangent to the orbit at every point, we see that a highly eccentric orbit is one for which **p** and **r** are nearly collinear over most of the orbit, making p small almost everywhere. The exceptions occur at the perigee (nearest point), where **r** and **p** are perpendicular but $|\mathbf{r}|$ is small, and apogee (farthest point), where **r** and **p** are again perpendicular but now $|\mathbf{p}|$ is small. Thus, we expect that $|\mathbf{L}|$ will be smaller for the more eccentric orbits. The extreme case is that for $L = 0$, where the classical orbit degenerates to a straight line.

Chapter 8

The quantum probabilities are found by integrating the radial probability density for each state, $P(r)$, from $r = 0$ to $r = a_0$. For the 2s state we find from Table 7.4 (with $Z = 1$ for hydrogen)

$$P_{2s}(r) = |rR_{2s}(r)|^2 = (8a_0)^{-1}(r/a_0)^2(2 - r/a_0)^2 e^{-r/a_0}$$

and

$$P = (8a_0)^{-1}\int_0^{a_0}(r/a_0)^2(2 - r/a_0)^2 e^{-r/a_0}\,dr$$

Changing variables from r to $z = r/a_0$ gives

$$P = 8^{-1}\int_0^1 (4z^2 - 4z^3 + z^4)e^{-z}\,dz$$

Repeated integration by parts gives

$$P = 8^{-1}\Big\{-(4z^2 - 4z^3 + z^4) - (8z - 12z^2 + 4z^3) - (8 - 24z + 12z^2) - (-24 + 24z) - (24)\Big\}e^{-z}\Big|_0^1$$

$$= 8^{-1}\Big\{-(1 + 0 - 4 + 0 + 24)e^{-1} + 8\Big\} = 0.034$$

For the 2p state of hydrogen

$$P_{2p}(r) = |rR_{2p}(r)|^2 = (24a_0)^{-1}(r/a_0)^4 e^{-r/a_0}$$

and

$$P = (24a_0)^{-1}\int_0^{a_0}(r/a_0)^4 e^{-r/a_0}\,dr = 24^{-1}\int_0^1 z^4 e^{-z}\,dz$$

Again integrating by parts, we get

$$P = 24^{-1}\left\{-z^4 - 4z^3 - 12z^2 - 24z - 24\right\}e^{-z}\Big|_0^1 = 24^{-1}\left\{-65e^{-1} + 24\right\}$$

$$= 0.0037$$

The probability for the 2p electron is nearly ten times smaller, suggesting that this electron spends more of its time in the outer regions of the atom where it is screened more effectively by the inner shell electrons.

Chapter 9

2.

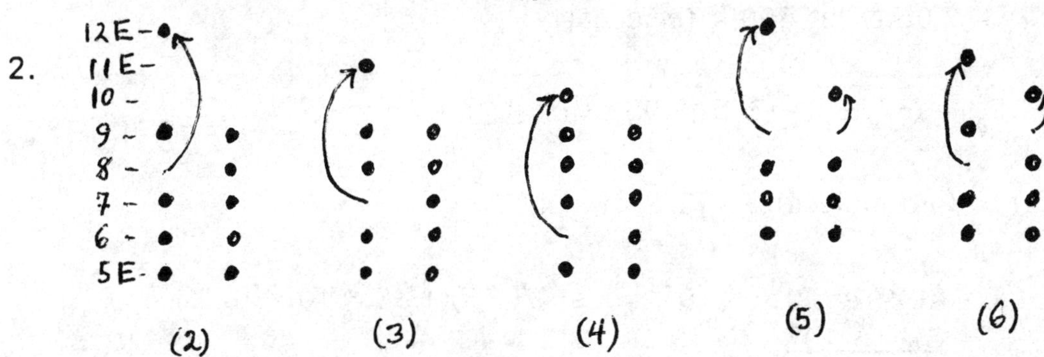

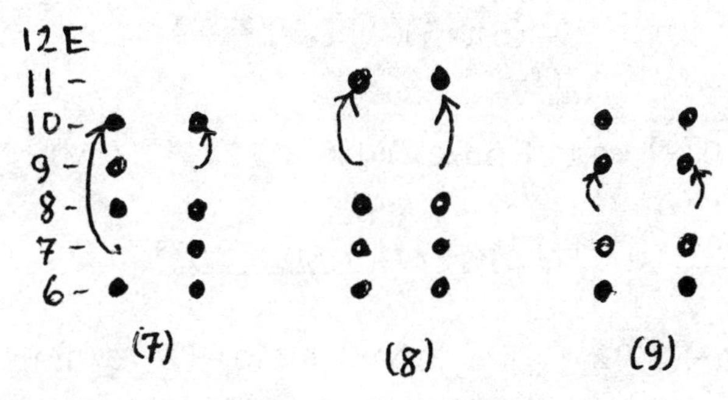

Thus n_{0E} = 1/9 × 2 + 1/9 × 2 + 1/9 × 2 + 1/9 × 2 + 1/9 × 2 + 1/9 × 2 + 1/9 × 2 + 1/9 × 2 + 1/9 × 2 = 2.00

n_{0E} through n_{5E} = 2.00

n_{6E} = 8(1/9 × 2) + (1/9 × 1) = 1.89

n_{7E} = 7(1/9 × 2) + (1/9 × 1) + (1/9 × 1) = 1.78

n_{8E} = 6(1/9 × 2) + (1/9 × 1) + (1/9 × 1) = 1.55

n_{9E} = 5(1/9 × 2) + (1/9 × 1) + (1/9 × 1) + (1/9 × 1) = 1.44

n_{10E} = (1/9 × 1) + (1/9 × 1) + (1/9 × 1) + (1/9 × 2) + (1/9 × 2) = 0.89

n_{11E} = (1/9 × 2) + (1/9 × 1) + (1/9 × 1) = 0.44

n_{12E} = (1/9 × 1) + (1/9 × 1) = 0.22

n_{13E} = (1/9 × 1) = 0.11

n_{14E} = 0.00

Minimum energy occurs for all levels filled up to 9E, corresponding to a total energy of 90E.

So E_F(0 K) = 9E

Chapter 9

3. (a) $E_F = 7.05$ eV at 300 K for copper.

$$E_{av} = (3/5)E_F = (3/5)(7.05 \text{ eV}) = \underline{4.23} \text{ eV}$$

(b) E_{av}(per molecule) = $(3/2)kT = 4.23$ eV

$$T = (4.23 \text{ eV})(1.6 \times 10^{-19} \text{ J/eV})/(1.5)(1.38 \times 10^{-23} \text{ J/K})$$
$$= \underline{3.27 \times 10^4} \text{ K}$$

---

4. Al: $E_F = 11.63$ eV

(a) $E_F = (h^2/2m)(3n/8\pi)^{2/3}$ or $n = (8\pi/3)(2mE_F/h^2)^{3/2}$ so

$$n = (8\pi/3)[(2)(9.11 \times 10^{-31} \text{ kg})(11.63 \text{ eV})(1.6 \times 10^{-19} \text{ J/eV})$$

$$\div (6.625 \times 10^{-34} \text{ J·s})^2]^{3/2} = \underline{1.80 \times 10^{29}} \text{ m}^{-3}$$

(b) $n' = \rho N_A/M = (2.7 \text{ g/cm}^3)(6.02 \times 10^{23} \text{ atoms/mole})/(27 \text{ g/mole})$

$$n' = 6.02 \times 10^{22} \text{ atoms/cm}^3 = 6.02 \times 10^{28} \text{ atoms/m}^3$$

$$\text{Valence} = n/n' = (18 \times 10^{28})/(6 \times 10^{28}) = \underline{3}$$

---

5. As $N_v = 4\pi N(m/2\pi kT)^{3/2} v^2 e^{-mv^2/2kT}$,

N_v has a maximum for $dN_v/dv = 0$. Therefore,

$$dN_v/dv = C[2ve^{-mv^2/2kT} + v^2(-2mv/2kT)e^{-mv^2/2kT}] \equiv 0$$

$$= Cve^{-mv^2/2kT}[2 - mv^2/kT] = 0$$

Thus we have extrema at $v = 0$, $v = \infty$, and

$$[2 - mv^2/kT] = 0 \quad \text{or} \quad v = (2kT/m)^{1/2}$$

As $N_v \to 0$ at $v = 0$ and ∞, $v = 0$ and $v = \infty$ are minima and $v = (2kT/m)^{1/2}$ is a maximum of N_v. Thus $v_{rms} = (2kT/m)^{1/2}$

Chapter 9

6. (a) A molecule moving with speed v takes d/v seconds to cross the cylinder, where d is the cylinder's diameter. In this time the detector rotates θ radians where $\theta = \omega \cdot t = \omega d/v$. This means the molecule strikes the curved glass plate at a distance from A of

$$S = (d/2)\theta = \omega d^2/2v. \qquad \text{As } m_{Bi_2} = 6.94 \times 10^{-22}\text{ g}$$

and

$$\langle v \rangle = \left[\frac{8kT}{\pi m}\right]^{1/2} = \left[\frac{(8)(1.38 \times 10^{-23}\text{ J/K})(850\text{ K})}{(\pi)(6.94 \times 10^{-25}\text{ kg})}\right]^{1/2} = 207\text{ m/s}$$

$$v_{rms} = (3kT/m)^{1/2} = 225\text{ m/s} \qquad v_{mp} = (2kT/m)^{1/2} = 184\text{ m/s},$$

$$S_{rms} = \left(\frac{6250 \times 2\pi}{60\text{ s}}\right)(0.10\text{ m})^2/(2)(225)\text{ m/s} = 1.45\text{ cm}$$

$$S_{\langle v \rangle} = \underline{1.58\text{ cm}} \qquad S_{mp} = \underline{1.78\text{ cm}}$$

---

7. [Strength of Absorption Line] = [Population of the initial state] · (Transition Probability]

Using M-B statistics to calculate the populations of E_1 and E_2 relative to the ground state, E_0, we find for the strength of absorption lines <u>at 4 K</u>:

$$S_{E_0 \to E_3} = (1)e^0 = 1.00 \qquad \text{at } E = 3\Delta$$

$$S_{E_1 \to E_3} = (2)e^{-\Delta/kT} = 2e^{\frac{-12.41 \times 10^{-5}\text{ eV}}{8.62 \times 10^{-5} \times 4\text{ eV}}} = 1.40 \qquad \text{at } E = 2\Delta$$

$$S_{E_2 \to E_3} = (2)e^{-2\Delta/kT} = 0.97 \qquad \text{at } E = 1\Delta$$

<u>At 1 K</u>,

$$S_{E_0 \to E_3} = 1.00 \qquad \text{at } E = 3\Delta$$

$$S_{E_1 \to E_3} = (2)e^{\frac{-12.41 \times 10^{-5}}{8.62 \times 10^{-5}}} = 0.47$$

$$S_{E_2 \to E_3} = (2)e^{\frac{-2(12.41 \times 10^{-5})}{8.62 \times 10^{-5}}} = 0.11$$

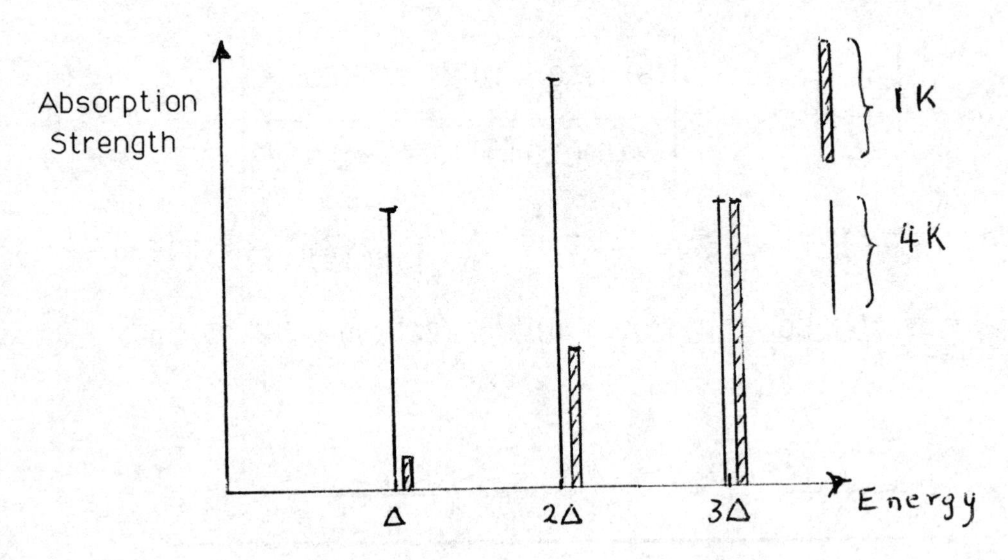

9. (a) Use Equation 9.10

$(N/V)\,h^3/(3kmT)^{3/2} << 1$ for M.B.

$N/V = (6.02 \times 10^{23})/(22.4\ m^3) = 2.69 \times 10^{22}\ m^{-3}$

$h^3 = 291 \times 10^{-102}\ (J{\cdot}s)^3$ $\qquad kT = 4.14 \times 10^{-21}\ J$

$m_{He} = (4\ g)/(6.02 \times 10^{23}) = 6.64 \times 10^{-27}\ kg$

So

$$\frac{(2.69 \times 10^{22}\ m^{-3})(2.91 \times 10^{-102}\ (J{\cdot}s)^3)}{(3 \times 4.14 \times 10^{-21}\ J \times 6.64 \times 10^{-27}\ kg)^{3/2}} = 1.04 \times 10^{-8}$$

Thus the MB distribution is a valid approximation for He gas at STP.

Chapter 9

9. (b) at 4 K,

$$N/V = 0.145/4\ (\text{moles/cm}^3) \times 6.02 \times 10^{23}\ \text{atoms/mole}$$

$$= 2.18 \times 10^{28}\ \text{atom/m}^3$$

and $kT = 5.52 \times 10^{-23}$ J so

$$\left(\frac{N}{V}\right)\frac{h^3}{(3kmT)^{3/2}} = \frac{(2.18 \times 10^{28}\ \text{m}^{-3})[291 \times 10^{-102}(\text{J·s})^3]}{(3 \times 5.52 \times 10^{-23}\ \text{J} \times 6.64 \times 10^{-27}\ \text{kg})^{3/2}} = 5.50$$

5.50 is not << 1 so that we must use the BE distribution to describe liquid He.

---

10. (a) $n_1 + n_2 = 10^{20}$

$$\frac{n_2}{n_1} = \exp\left(\frac{-4.86\ \text{eV} \times 1.602 \times 10^{-19}\ \text{J/eV}}{1.38 \times 10^{-23}\ \text{J/K} \times 1.600 \times 10^3\ \text{K}}\right) = 4.98 \times 10^{-16}$$

Assuming $n_1 \approx 10^{20}$,

$$n_2 \approx n_1 (4.98 \times 10^{-16}) = (10^{20})(4.98 \times 10^{-16}) = 4.98 \times 10^4$$

(b) Power emitted = number of photons emitted/s × (energy/photon)

$$= (1/\tau) \times n_2 \times 4.86\ \text{eV}$$

$$= 10^7\ \text{s}^{-1} \times 4.98 \times 10^4 \times 4.86\ \text{eV} \times 1.602 \times 10^{-19}\ \text{J/eV}$$

$$= 3.88 \times 10^{-7}\ \text{J/s}$$

$$= \underline{0.388}\ \mu\text{W}$$

Chapter 9

11.

$$\bar{E}_{photon} = \frac{\int_0^\infty E F_{BE}(E)\, g(E)\, dE}{N/V}$$

$$= \frac{\int_0^\infty E F_{BE}(E)\, g(E)\, dE}{\int_0^\infty F_{BE}(E)\, g(E)\, dE}$$

$$= \frac{\int_0^\infty \frac{8\pi E^3}{(hc)^3} \frac{1}{(e^{E/kT}-1)}\, dE}{\int_0^\infty \frac{8\pi E^2}{(hc)^3} \frac{1}{(e^{E/kT}-1)}\, dE}$$

$$= \frac{\frac{8\pi}{(hc)^3} (kT)^4 \int_0^\infty \frac{(E/kT)^3 (dE/kT)}{(e^{E/kT}-1)}}{\frac{8\pi}{(hc)^3} (kT)^3 \int_0^\infty \frac{(E/kT)^2 (dE/kT)}{(e^{E/kT}-1)}}$$

$$\bar{E}_{photon} = \frac{kT \int_0^\infty \frac{z^3\,dz}{e^z - 1}}{\int_0^\infty \frac{z^2\,dz}{e^z - 1}} = \frac{kT\,\pi^4/15}{2.41}$$

(b) The average energy per photon at 6000 K is

$$\bar{E} = \frac{kT\,\pi^4}{(15)(2.41)} = \frac{(8.62 \times 10^{-5}\ \text{eV/K})(6 \times 10^3\ \text{K})\pi^4}{36.15}$$

$$= \underline{1.39}\ \text{eV}$$

---

12. $$E_{av} = (1/n)\int_0^\infty E\,N(E)\,dE$$

but at $T = 0$, $E = 0$ for $E > E_F$ and we can take $N(E) = CE^{1/2}$

Since $f(E) = 1$ for $E < E_F$ and $f(E) = 0$ for $E > E_F$.

$$E_{av} = (1/n)\int_0^{E_F} CE^{3/2}\,dE = (C/n)\int_0^{E_F} E^{3/2}\,dE = (2C/5n)E_F^{5/2}$$

But $C/n = (3/2)E_F^{-3/2}$, so that

$$E_{av} = (2/5)(3/2)E_F^{-3/2}E_F^{5/2} = (3/5)E_F$$

---

13. Equation 9.31 gives $E_F(0)$ in terms of N/V as

$$E_F = \left(\frac{h^2}{2m}\right)\left(\frac{3N}{8\pi V}\right)^{2/3}$$

Substituting the mass of a proton, and noting that A = 64 for Zn

$m = 1.67 \times 10^{-27}$ kg ; $N = A/2 = 32$

and $V = (4/3)\pi R^3 = (4/3)(\pi)(4.8 \times 10^{-15} \text{ m})^3 = 4.6 \times 10^{-43} \text{ m}^3$

yields

$$E_F = \frac{(6.62 \times 10^{-34})^2 \text{ J}^2\text{s}^2}{3.34 \times 10{-27} \text{ kg}} \times \left(\frac{3 \cdot 32}{8\pi \cdot 4.6 \times 10^{-43} \text{ m}^3}\right)^{2/3}$$

$$= 5.3 \times 10^{-12} \text{ J} = 33.4 \text{ MeV}$$

$$E_{av} = (3/5)E_F = \underline{20} \text{ MeV}$$

These energies are of the correct order of magnitude for nuclear particles.

---

14. $E_F = (h^2/2m)(3n/8\pi)^{2/3}$

$$E_F = (6.625 \times 10^{-34} \text{ J}\cdot\text{s})^2/(2 \times 9.11 \times 10^{-31} \text{ kg} \times 1.6 \times 10^{-19} \text{ J/ev})]$$

$$\times (3/8\pi)^{2/3} n^{2/3}$$

$$E_F = (3.65 \times 10^{-19}) n^{2/3} \text{ eV}$$

with n measured in electrons/m^3.

Chapter 9

15. $f = [e^{(E - E_f)/kT} + 1]^{-1}$; $E_F = 7.05$ ev ;

$kT = (1.38 \times 10^{-23}\ J/K)(300\ K) = 4.14 \times 10^{-21}\ J = 0.0259\ eV$

At $E = 0.99E_F$

$f = [e^{-0.01E_f/kT} + 1]^{-1} = 1/(e^{-0.0705/0.0259} + 1) = 1/1.0657 = \underline{0.938}$

thus 93.8% probability

---

16. $f = [e^{(E-E_F)/kT} + 1]^{-1}$. When $E = E_F$

$f = (e^{\circ} + 1)^{-1} = 1/2 = \underline{0.50}$ or $\underline{50\%}$ probability (independent of T!)

---

17. $\rho = 0.971\ g/cm^3$, $M = 23.0$ g/mole (sodium)

(a) $n = N_A\rho/M$

$n = (6.02 \times 10^{23}$ electrons/mole$)(0.971\ g/cm^3)(23.0$ g/mole$)$

$n = 2.54 \times 10^{22}$ electrons/cm^3 = $\underline{2.54 \times 10^{28}}$ electrons/m^3

(b) $E_F = (h^2/2m)(3n/8\pi)^{2/3}$

$E_F = [(6.625 \times 10^{-34})^2/(2 \times 9.11 \times 10^{-31})][(3 \times 2.54 \times 10^{28})/8\pi]^{2/3}$

$E_F = 5.04 \times 10^{-19}\ J = \underline{3.15}\ eV$

(c) $v_F = (2E_F/m)^{1/2} = [2 \times 5.04 \times 10^{-19}\ J/(9.11 \times 10^{-31}\ kg)]^{1/2}$

$v_F = \underline{1.05 \times 10^6}$ m/s

---

18. Taking $E_F = 5.48$ eV for sodium at 800 K we have

$f = [e^{(E-E_F)/kT} + 1]^{-1} = 0.95$

$e^{(E-E_F)/kT} = (1/0.95) - 1 = 0.05263$

$(E - E_F)/kT = \ln(0.05263) = -2.944$

$E - E_F = -2.944(1.38 \times 10^{-23} \times 800 \text{ J})/(1.6 \times 10^{-19} \text{ J/eV})$

$E - E_F = -0.203$ eV or $E = \underline{5.28}$ eV

---

19. $d = 1 \text{ mm} = 10^{-3} \text{ m}$; $V = (10^{-3} \text{ m})^3 = 10^{-9} \text{ m}^3$

The density of states $= g(E) = CE^{1/2} = \{8(2)^{1/2}\pi m^{3/2}/h^3\}\, E^{1/2}$

$g(E) = 8(2)^{1/2}\pi(9.11 \times 10^{-31} \text{ kg})^{3/2}$

$\times [(4.0 \text{ eV})(1.6 \times 10^{-19} \text{ J/eV})]^{1/2}/(6.626 \times 10^{-34} \text{ J·s})^3$

$g(E) = (8.50 \times 10^{46}) \text{ m}^{-3}\text{J}^{-1} = (1.36 \times 10^{28}) \text{ m}^{-3}\text{eV}^{-1}$

So the total number of electrons $= N = g(E)(\Delta E)V$ or

$N = (1.36 \times 10^{28} \text{ m}^{-3} \text{ eV}^{-1})(0.025 \text{ eV})(10^{-9} \text{ m}^3) = \underline{3.40 \times 10^{17}}$ electrons

---

20. (a) The density of states at the energy E is $g(E) = CE^{1/2}$

Hence, the ratio of the number of allowed levels at 8.5 eV to the number of allowed levels at 7.0 eV is

$g(8.5)/g(7.0) = C(8.5)^{1/2}/C(7.0)^{1/2} = (8.5/7.0)^{1/2} = \underline{1.1}$

20. (b) The number of electrons occupying an energy level E is

$$N(E) \propto E^{1/2}/[e^{(E - E_F)/kT} + 1],$$

hence the ratio required is

$$N(8.5)/N(7.0) = (8.5/7.0)^{1/2}(e^{0} + 1)/[e^{(8.5-7.0)/kT} + 1]$$

$$= (1.1)[2/(e^{1.5/0.0259} + 1)] = 1.55 \times 10^{-25}$$

Comparing this with (a), we see that very few states with $E > E_F$ are occupied.

Chapter 10

1. We follow the development in Equations 10.1 through 10.12 now taking the degeneracies of $n_2(g_2)$ and $n_1(g_1)$ explicitly into account.

$$N_2 = N_1(g_2/g_1)e^{-(E_2-E_1)/kT} = N_1(g_2/g_1)\, e^{-hf/kT} \qquad (1)$$

Invoking equal numbers of transitions up and down per unit time:

$$N_1 u(f,T)\, B_{12} = N_2[B_{21}u(f,T) + A_{21}] \qquad (2)$$

Substituting (1) into (2) gives

$$N_1 u(f,T) B_{12} = N_1 g_2/g_1\, e^{-hf/kT}[B_{21}u(f,T) + A_{21}]$$

$$u\left[B_{12} - g_2/g_1\, e^{-hf/kT}B_{21}\right] = (g_2/g_1)\, e^{-hf/kT} A_{21}$$

or

$$u(f/T) = \frac{A_{21}}{(g_1/g_2)B_{12}\, e^{hf/kT} - B_{21}} = \frac{A_{21}}{B_{21}(e^{hf/kT}-1)}$$

where $\qquad B_{21} = (g_1/g_2)B_{12}$

Comparing to the Planck expression $u(f,T) = (8\pi hf^3/c^3)[e^{hf/kT} - 1]^{-1}$ we find

$$A_{21}/B_{21} = 8\pi hf^3/c^3$$

or $\qquad B_{21} = (A_{21}c^3)/8\pi hf^3$

and $\qquad B_{12} = (g_2/g_1)A_{21}c^3/8\pi hf^3$

The induced probability of emission per unit time per atom is

$$W_i(2 \rightarrow 1) = B_{21}u(f,T) = \left[A_{21}c^3/8\pi hf^3\right]u(f,T)$$

and of absorption is

$$W_i(1 \rightarrow 2) = B_{12}u(f,T) = (c^3/8\pi hf^3)(g_2/g_1)A_{21}u(f,T)$$

Thus the net gain in intensity of a cavity mode with energy hf is

$$(dI/dt)_{gain} = \left[N_2W_i(2 \rightarrow 1) - N_1W_i(1 \rightarrow 2)\right]hfc/V$$

and substituting for $W_i(2 \rightarrow 1)$ and $W_i(1 \rightarrow 2)$

$$(dI/dt)_{gain} = (n_2 - n_1 g_2/g_1)hfcW_i$$

where $$W_i = [c^3A_{21}/8\pi hf^3]u(f,T) = [c^2/8\pi hf^3t_s]g(f)I(f)$$

in agreement with Equation 10.12.

---

2. (a) $n(E_2)/n(0)$ at 300 K $= e^{-0.5505\text{ eV}/0.0258\text{ eV}} = 5.41 \times 10^{-10}$

$n(E_1)/n(0)$ at 300 K $= e^{-0.0755\text{ eV}/0.0258\text{ eV}} = 0.0535$

Thus about 5% of the Uranium ions are in the E_1 state at 300 K.

(b) $n(E_2)/n(0)$ at 77 K $= e^{-0.5505/0.00663} = 8.71 \times 10^{-37}$

$n(E_1)/n(0)$ at 77 K $= e^{-0.0755/0.00663} = 1.13 \times 10^{-5}$

At 77 K the populations of both E_2 and E_1 are ≈ 0 and the system behaves like a 4-level laser with a terminal level widely separated ($\approx 12kT$) from the ground state.

(c) Using $t_p = (1/\alpha)(L/c)$, gives $t_p = 2.8 \times 10^{-9}$ s

Using $\Delta N_c/V = 4\pi^2f^2\Delta f t_s/c^3t_p$ with

$f = 3827 \times 3 \times 10^{10} = 1.148 \times 10^{14}$ Hz

and

$\Delta f = 20\text{ cm}^{-1} \times 3 \times 10^{10}\text{ cm/s} = 6.0 \times 10^{11}$ Hz yields

2. (Cont'd) $\Delta N_c/V = 1.4 \times 10^{15}\ \text{cm}^{-3}$

The minimum pumping power is

$$P = \frac{4\pi^2 h f^3 \Delta f}{c^3 t_p} = 0.81\ \text{W/cm}^3$$

(d) $1/\lambda = 3827\ \text{cm}^{-1}$ so $\lambda = 1/(3827\ \text{cm}^{-1}) = 2.61 \times 10^{-4}\ \text{cm}$

$= 2.61 \times 10^{-6}\ \text{m} = 2610\ \text{nm}$ (in the IR region)

(e) The pumping power for Ca F:U^{3+} laser is less because

(1) Ruby emits higher energy light

(2) Ca F:U^{3+} is a 4-level laser and ruby is a 3-level laser

---

4. If laser operation is confined to a single mode, than $\Delta\lambda$, the laser linewidth, equals, at most, the change in wavelength between adjacent modes.

$$\Delta\lambda = \lambda^2/2L(n - \lambda\, dn/dx) \qquad (10.29)$$

so $\Delta\lambda = (694.4\ \text{nm})^2/(20\ \text{cm})(10^7\ \text{nm/cm})(1.8) = 0.00133\ \text{nm}$

and $\lambda_{laser} = (694.4 \pm 0.001)\ \text{nm}$

Thus the condition $m(\lambda/2n) = L$ restricts the laser wavelength to about 1/200 of the thermal line width!

---

5. (a) Since kT at 300 K = 0.0258 eV, and

$$2000\ cm^{-1} = 2000\ cm^{-1} \times 1.241\ eV/cm^{-1} = 0.2482\ eV,$$

the laser terminal level and the ground state are separated by about 10kT at room temperature. Hence, there will be very little population of the laser terminal level and a population inversion will be easy to set up with little pumping.

$$\frac{n(\text{laser terminal level})}{n(\text{ground state})} \approx \frac{e^{-0.2482/0.0258}}{e^{0}} = 6.6 \times 10^{-5}$$

Since less than 0.01% of the atoms will be in the laser terminal level at 300 K, this laser acts like a 4-level system.

(b) $t_p = (1/\alpha)(L/c') = (1/\alpha)(nL/c)$

$$= (1/0.07)\ (1.8)(5\ cm)/(3 \times 10^{10}\ cm/s)$$

$$= 4.28 \times 10^{-9}\ s$$

$$f = (1/\lambda)c = (9398.50\ cm^{-1})(3.00 \times 10^{10}\ cm/s) = 2.819 \times 10^{14}\ Hz$$

$$\Delta f = (1/\lambda)c = (15\ cm^{-1})(3 \times 10^{10}\ cm/s) = 4.5 \times 10^{11}\ Hz$$

Equation 10.17, $\Delta N_c/V = (4\pi^2) f^2 \Delta f / c'^3 (t_s/t_p)$ becomes

$$\Delta N_c/V = \frac{(4\pi^2)(2.819 \times 10^{14})^2\ s^{-2}\ (4.5 \times 10^{11}\ s^{-1})}{(1.67 \times 10^{10}\ cm/s)^3}\left(\frac{10^{-4}}{4.28 \times 10^{-9}}\right)$$

$$= 7.1 \times 10^{15}\ \text{ions}/cm^3$$

(Note that c' is the velocity of light in the laser material in Eq. 10.17).

5. (c) $(P = \Delta N_C/V\ (hf/t_S)$

$$= [(7.1 \times 10^{15})/\text{cm}^3\ ](6.63 \times 10^{-34}\ \text{J}\cdot\text{s})(2.82 \times 10^{14}\ \text{s}^{-1})/10^{-4}\ \text{s}$$

$P = \underline{13\ \text{W/cm}^3}$

(d) The threshold pumping power is generally observed to be greater for rare earth ions in glass matrices than in crystal matrices. This occurs because of the more random distribution of atoms in glasses which in turn leads to broader absorption lines and higher light losses than in pure single crystals.

---

8. (a) From Example 10.2

$$\Delta N_C/V = n(E) - n(^4A_2)\ g(E)/g(^4A_2) = 5.2 \times 10^{16}\ \text{ions/cm}^3$$

or

$$n(^4A_2) = 2n(E) - 1.02 \times 10^{17} \qquad (1)$$

The ratio of 2A and E populations is given by the Maxwell-Boltzmann relation

$$n(2A)/n(E) = [g(2A)/g(E)]\, e^{-E_{2A}/kT} \Big/ e^{-E_E/kT}$$

Using $E_{2A} - E_E = (29\ \text{cm}^{-1})(1.24 \times 10^{-4}\ \text{eV/cm}^{-1}) = 0.00360\ \text{eV}$

and kT at room temperature = 0.0259 eV gives

$$n(2A) = 0.87n(E) \qquad (2)$$

Assuming that all of the Cr ions are in the three lowest levels we have

$$n(2A) + n(E) + n(^4A_2) = N = 2 \times 10^{19}\ \text{ions/cm}^3$$

or substituting (1) and (2) into this equation gives

$$0.87\ n(E) + n(E) + 2\ n(E) - 1.02 \times 10^{17} = 2.00 \times 10^{19}$$

8. (Cont'd) Finally we obtain

$n(E) = 0.52 \times 10^{19}$ Cr ions/cm^3
$n(2A) = 0.45 \times 10^{19}$ Cr ions/cm^3
$n(^4A_2) = 1.02 \times 10^{19}$ Cr iona/cm^3

(b) $P_{min} \approx [n(E) + n(2A)]\cdot hf/t_s$

$$P_{min} = \frac{[(0.45 + 0.52) \times 10^{19}\ \text{cm}^{-3}](6.63 \times 10^{-34}\ \text{J}\cdot\text{s})(4.32 \times 10^{14}\ \text{s}^{-1})}{3 \times 10^{-3}\ \text{s}}$$

$= \underline{930\ \text{W/cm}^3}$

---

9. (a) $dm/d\lambda = d/d\lambda\{2Ln/\lambda\} = (2L/\lambda)(dn/d\lambda - n/\lambda)$

Replacing dm and $d\lambda$ with Δm and $\Delta\lambda$ yields

$$\Delta\lambda = (\lambda^2 \Delta m)/2L(\lambda\ dn/d\lambda - n)$$

or $$|\Delta\lambda| = \lambda^2/2L(n - \lambda\ dn/d\lambda)$$

Since $\Delta\lambda$ is negative for $\Delta m = +1$.

(b) $|\Delta\lambda| = (837 \times 10^{-9}\ \text{m})^2 \div$

$(0.6 \times 10^{-3}\ \text{m})\ [3.58 + (837\ \text{nm})\ (3.8 \times 10^{-4}\ \text{nm}^{-1})]$

$= 3.0 \times 10^{-10}\ \text{m} = 0.3\ \text{nm}$

(c) $|\Delta\lambda| = (633 \times 10^{-9}\ \text{m})^2 / (0.6 \times 10^0\ \text{m})(1)$
$= 6.7 \times 10^{-13}\ \text{m} = 0.00067\ \text{nm}$

The controlling factor is <u>cavity length</u>, L.

Chapter 11

1. $f = (1/2\pi)(K/\mu)^{1/2}$ and $5.63 \times 10^{13} = (1/2\pi)(1530/\mu)^{1/2}$

giving $\mu = 12.2 \times 10^{-27}$ kg. From Equation 11.3,

$\mu = 14(16)/(14 + 16) = 7.47\ u = \underline{12.4 \times 10^{-27}}$ kg.

---

2. For the $\ell = 1$ to $\ell = 2$ transition, $\Delta E = hf = [2(2 + 1) - 1(1 + 1)]\hbar^2/2I$

or $hf = 2\hbar^2/I$; Solving for I gives

$I = 2\hbar^2/hf = h/2\pi^2 f = 6.626 \times 10^{-34}/2\pi^2(2.30 \times 10^{11})$

$= 1.46 \times 10^{-46}$ kg·m^2; $\mu = m_1 m_2/(m_1 + m_2) = 1.14 \times 10^{-26}$ kg

$r = (I/\mu)^{1/2} = \underline{1.13\ \text{Å}}$, no change .

---

3. (a) HI:

$$1(127)/128 = \mu\ ;\quad f = (1/2\pi)(K/\mu_1)^{1/2}$$

$A^2 = (h/2\pi)(1/K\mu)^{1/2}$

$= [(6.626 \times 10^{-34})/2\pi][1/(320)(127/128)(1.66 \times 10^{-27})]^{1/2}$

$= 1.45 \times 10^{-22}$

$A = 1.20 \times 10^{-11}$ m $= 0.120$ **Å** $= \underline{0.0120 \text{ nm}}$

(b) HF ; $\mu = 1(19)/20$

$A^2 = (6.626 \times 10^{-34})/2\pi\ [1/970(19/20)(1.66 \times 10^{-27})]^{1/2}$

$= 8.53 \times 10^{-23}$

$A = 9.23 \times 10^{-12}$ m $= 0.0923$ **Å** $= \underline{0.00923 \text{ nm}}$

---

4. (a) The separation between two adjacent rotational levels is given by

$$\Delta E = (\hbar^2/I)J', \quad \text{therefore} \quad \Delta E_{1,0} = \Delta E_{6,5}/6$$

$$\lambda_{1,0} = 6\lambda_{6,5} = 6(1.35 \text{ cm}) = 8.10 \text{ cm}$$

$$f_{1,0} = c/\lambda_{1,0} = (3 \times 10^{10} \text{ cm/s})/(8.1 \text{ cm}) = \underline{3.70 \text{ GHz}}$$

(b) $\Delta E_{10} = hf_{10} = \hbar^2/I$;

$$I = \hbar/2\pi f_{10} = (6.63 \times 10^{-34} \text{ J·s})/(2\pi)^2(3.7 \times 10^9 \text{ Hz})$$

$$I = \underline{4.53 \times 10^{-45}} \text{ kg·m}^2$$

---

5. HF molecule : $r_0 = 0.92$ Å

(a) $\mu = m_1m_2/m_1 + m_2 = (1)(19)/(1 + 19)$ u $= \underline{0.95}$ u

(b)

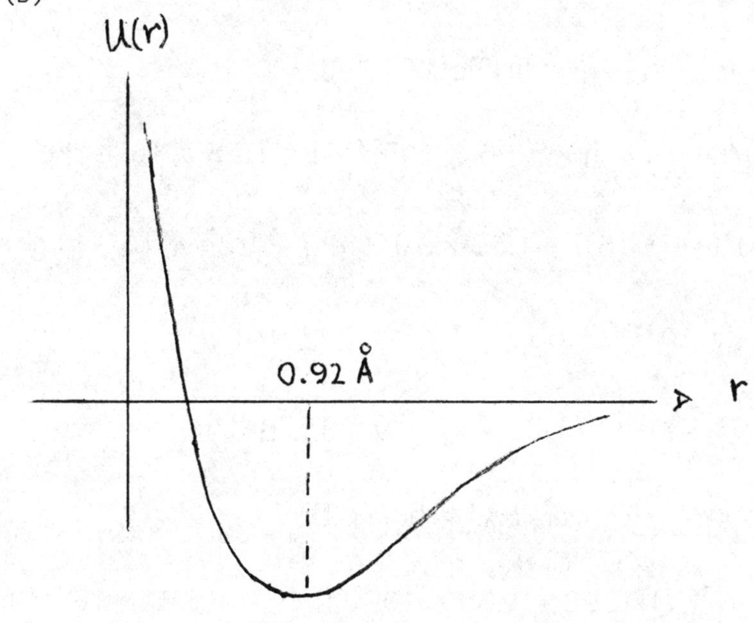

Chapter 11

6. $HC\ell$ molecule in the $\ell = 1$ rotational energy level :

$r_0 = 1.275$ **Å** $\qquad E_{rot} = (\hbar^2/2\ I)\ell(\ell + 1)$

For $\ell = 1$, $E_{rot} = \hbar^2/I = I\omega^2/2$

$$\omega = (2\hbar^2/I^2)^{1/2} = (\hbar/I)\sqrt{2}$$

$I = [m_1 m_2/(m_1 + m_2)]^{1/2} r_0^2 = [1(35)/(1 + 35)\ u] r_0^2$

$= [0.9722\ u \times 1.66 \times 10^{-27}\ kg/u] \times (1.275 \times 10^{-10}\ m)^2$

$= 2.62 \times 10^{-47}\ kg{\cdot}m^2$ $\qquad$ Therefore,

$\omega = (\hbar/I)\sqrt{2} = [(1.055 \times 10^{-34}\ J{\cdot}s)/(2.62 \times 10^{-47}\ kg{\cdot}m^2)]\sqrt{2}$

$= \underline{5.69 \times 10}^{12}$ rad/s

---

7. (a) HI $\quad \Delta E_{0\to1} = hf = (h/2\pi)(K/\mu)^{1/2}$ or $2\pi f = (K/\mu)^{1/2}$ or

$4\pi^2 f^2 \mu = K$ where $\mu_{HI} = m_1 m_2/(m_1 + m_2)$

$\mu_{HI} = [1(127)/(1 + 127)] \times 1.66 \times 10^{-27}\ kg = 1.65 \times 10^{-27}\ kg$

$\mu_{NO} = [14(16)/(14 + 16)] \times 1.66 \times 10^{-27}\ kg = 1.24 \times 10^{-26}\ kg$

so $K_{HI} = 4\pi^2(6.69 \times 10^{13})^2(1.65 \times 10^{-27}) = \underline{292}$ N/m

$K_{NO} = 4\pi^2(5.63 \times 10^{13})^2(1.24 \times 10^{-26}) = \underline{1550}$ N/m

(b) $KA^2/2 = (\hbar/2)(K/\mu)^{1/2}$ or $A^2 = \hbar(1/K\mu)^{1/2}$

$A^2_{HI} = (1.06 \times 10^{-34})[1/(292 \times 1.65 \times 10^{-27})]^{1/2} = 1.519 \times 10^{-22}\ m^2$

$A_{HI} = \underline{0.123}$ **Å** $= \underline{0.0123}$ nm

Likewise, A_{NO} = $\underline{0.0492}$ Å = $\underline{0.00492}$ nm ;
N and O are "cemented" by more electrons.

---

8. $\mu = m_1m_2/(m_1 + m_2) = (1)(35)/(1 + 35) = (35/36)$ u

$\mu = (35/36)(1.67 \times 10^{-27})$ kg $= 1.62 \times 10^{-27}$ kg

(a) $I = \mu r^2 = (1.62 \times 10^{-27}$ kg$)(1.28 \times 10^{-10}$ m$)^2 = 2.65 \times 10^{-47}$ kg·m^2

$E_{rot} = (\hbar^2/2I)\, \ell(\ell + 1)$

$\hbar^2/2I = (1.054 \times 10^{-34})^2/(2 \times 2.65 \times 10^{-47})$ J $= 2.1 \times 10^{-22}$ J

$= 1.31 \times 10^{-3}$ eV

$E_{rot} = (1.31 \times 10^{-3}$ eV$)\, \ell(\ell + 1)$

$\ell = 0$ $E_{rot} = \underline{0}$
$\ell = 1$ $E_{rot} = \underline{2.62 \times 10^{-3}}$ eV
$\ell = 2$ $E_{rot} = \underline{7.86 \times 10^{-3}}$ eV
$\ell = 3$ $E_{rot} = \underline{1.57 \times 10^{-3}}$ eV

(b) $U = Kx^2/2$ $U = 0.15$ eV when $x = 0.1$ Å

$(0.15$ eV$)(1.6 \times 10^{-19}$ J/eV$) = K(10^{-11}$ m$)^2/2$

$K = \underline{480}$ N/m

$f = (1/2\pi)(K/\mu)^{1/2} = (1/2\pi)[480/(1.62 \times 10^{-27})]^{1/2} = \underline{8.66 \times 10^{13}}$ Hz

(c) $E_{vib} = (\upsilon + 1/2)hf$

$hf = (6.63 \times 10^{-34})(8.66 \times 10^{13})$ J $= 5.74 \times 10^{-20}$ J $= \underline{0.359}$ eV

$E_0 = hf/2 = 2.87 \times 10^{-20}$ J $= \underline{0.179}$ eV

$E = KA^2/2$; $2.87 \times 10^{-20}\ J = (480\ N/m)A_0{}^2/2$

$A_0 = (2E/K)^{1/2} = 1.09 \times 10^{-11}\ m = \underline{0.109}\ \text{Å} = \underline{0.0109}\ nm$

$E_1 = (3/2)hf = 8.61 \times 10^{-20}\ J = \underline{0.538}\ eV$

$A_1 = (2E/K)^{1/2} = 1.89 \times 10^{-11}\ m = \underline{0.189}\ \text{Å} = \underline{0.0189}\ nm$

(d) $hc/\lambda_{max} = \Delta E_{min}$ $\lambda_{max} = hc/\Delta E_{min}$

Rotational

$$\Delta E_{min} = E_{J'=1} - E_{J'=0} = 2.62 \times 10^{-3}\ eV$$

$$hc = 12400\ eV \cdot \text{Å}$$

$\lambda_{max} = 12400/(2.62 \times 10^{-3}) = \underline{4.73 \times 10^6}\ \text{Å}$

$\lambda_{max} = 4.73 \times 10^{-4}\ m$ (microwave range).

Vibrational

$$\Delta E_{min} = hf$$

$\lambda_{max} = hc/\Delta E_{min} = hc/hf = c/f = (3 \times 10^8\ m/s)/(8.66 \times 10^{13}\ Hz)$

$\lambda_{max} = \underline{3.46 \times 10^{-6}}\ m = \underline{3.46\ \mu m}$ (infrared range).

---

9. We take $E_{vib} = 4.5$ eV in the formula $E_{vib} = (\upsilon + 1/2)\hbar\omega$ to get

$$4.5\ eV = (\upsilon + 1/2)(6.582 \times 10^{-16}\ eV \cdot s)(8.277 \times 10^{14}\ rad/s)$$

$$= (\upsilon + 1/2)(0.5448\ eV) \quad \text{or} \quad \upsilon = 7.760$$

Of course, υ must be an integer, so that $\upsilon = 7$ represents the highest vibrational level that can be excited without the molecule coming apart.

---

10. The angular momentum of this system is

$$L = mvR_o/2 + mvR_o/2 = mvR_o$$

According to Bohr theory, L must be a multiple of $\hbar$, $L = mvR_o = n\hbar$, or

$$v = n\hbar/mR_o \qquad \text{with} \qquad n = 1, 2, \ldots$$

The energy of rotation is then

$$E_{rot} = (1/2)mv^2 + (1/2)mv^2 = m(n\hbar/mR_o)^2$$

$$= n^2/\hbar^2/mR_o^2 , \qquad n = 1, 2, \ldots$$

From Equation 11.5 the allowed energies of rotation are

$$E_{rot} = (\hbar^2/2I_{cm})\{\ell(\ell + 1)\}, \qquad \ell = 0, 1, 2, \ldots$$

where I_{cm} is the moment of inertia about the center of mass. In the present case, we have

$$I_{cm} = m(R_o/2)^2 + m(R_o/2)^2 = mR_o^2/2$$

Thus,

$$E_{rot} = (\hbar^2/mR_o^2)\{\ell(\ell + 1)\} \qquad \ell = 0, 1, 2, \ldots$$

We see that $\ell(\ell + 1)$ replaces n^2 in the Bohr result. The two are indistinguishable for large quantum numbers (Correspondence Principle), but disagree markedly when n (or ℓ) is small. In particular, E_{rot} can be zero according to Quantum Mechanics, while the minimum rotational energy in the Bohr theory is $\hbar^2/mR_o^2$ for $n = 1$.

Chapter 11

11. The excitation energy for rotation is the smallest nonzero energy given by Equation 11.5, and corresponds to $\ell = 1$:

$$\Delta E_{rot} = \hbar^2/I_{cm}$$

with I_{cm} now the moment of inertia about the internuclear line. Now the moment of inertia of a sphere about any axis passing through its center is

$$I = (2/5)mr^2$$

in terms of its mass m and radius r. Thus, for our diatomic molecule, we have

$$I_{cm} = (2)(2/5)mr^2 = (4/5)mr^2$$

and

$$\Delta E_{rot} = (5/4)\hbar^2/mr^2$$

Using r = 10 fm and m = 938.28 MeV/c^2 for the proton gives

$$\Delta E_{rot} = (5/4)(197.3 \text{ MeV}\cdot\text{fm}/c)^2/[(938.28 \text{ MeV}/c^2)(10 \text{ fm})^2]$$

$$= \underline{0.519 \text{ MeV}}$$

---

12. At the equilibrium separation R, U_{eff} is a minimum:

$$0 = dU_{eff}/dr\,\Big|_R = \mu\omega_0^2(R - R_0) - \ell(\ell + 1)\hbar^2/\mu R^3$$

or

$$R = R_0 + [\ell(\ell + 1)\hbar^2/\mu^2\omega_0^2](1/R^3)$$

For $\ell << \mu\omega_0 R_0^2/\hbar$, the second term on the right represents a small correction, and may be approximated by substituting for R its approximate value R_0 to get the next approximation

$$R \approx R_0 + [\ell(\ell + 1)\hbar^2/\mu^2\omega_0^2](1/R_0^3)$$

The value of U_{eff} at R is the energy offset U_0:

$$U_0 = U_{eff}(R) = (1/2)\mu\omega_0^2[\ell(\ell + 1)\hbar^2/\mu^2\omega_0^2R^3]^2 + \ell(\ell + 1)\hbar^2/2\mu R^2$$

$$= [\ell(\ell + 1)\hbar^2/2\mu R^2]\{\ell(\ell + 1)\hbar^2/\mu^2\omega_0^2R^4 + 1\}$$

$$\approx \ell(\ell + 1)\hbar^2/2\mu R_0^2$$

The curvature at the new equilibrium point is

$$d^2U_{eff}/dr^2|_R = \mu\omega_0^2 + 3\ell(\ell + 1)\hbar^2/\mu R^4$$

and is identified with $\mu\omega_\ell^2$ to get the corrected oscillator frequency

$$\omega_\ell^2 = \omega_0^2 + 3\ell(\ell + 1)\hbar^2/\mu^2R^4$$

$$\approx \omega_0^2 + 3\ell(\ell + 1)\hbar^2/\mu^2R_0^4$$

Since the second term on the right is small by assumption, ω_ℓ differs little from ω_0, so that we may write

$$\omega_\ell^2 - \omega_0^2 = (\omega_\ell - \omega_0)(\omega_\ell + \omega_0) \approx 2\omega_0\Delta\omega$$

The fractional change in frequency is then

$$\Delta\omega/\omega_0 \approx 3\ell(\ell + 1)\hbar^2/2\mu^2\omega_0^2R_0^4$$

---

13. The condition for equilibrium is

$$0 = dU/dr = 2U_0\{1 - e^{-\alpha(r - R_0)}\}\{\alpha e^{-\alpha(r - R_0)}\}$$

from which it is clear that $r = R_0$ is the only finite solution. The potential energy at equilibrium is

$$U(R_0) = U_0\left\{1 - e^{-\alpha(0)}\right\}^2 - U_0 = -U_0$$

Near equilibrium, $r - R_0$ is small and we can appeal to the expansion $e^x = 1 + x + \ldots$ to write approximately

$$1 - e^{-\alpha(r - R_0)} \approx + \alpha(r - R_0)$$

and

$$U(r) \approx U_0\{\alpha(r - R_0)\}^2 - U_0$$

This has the harmonic form

$$U(r) = (1/2)K(r - R_0)^2 + \text{ constant}$$

with $K = 2U_0\alpha^2$.

From the lowest vibrational level

$$E_{vib} = -U_0 + (1/2)\hbar\omega - (\hbar\omega)^2/16U_0$$

an energy of $-E_{vib}$ must be supplied to separate the molecule into its constituent atoms. Thus, the dissociation energy must be

$$E_{diss} = U_0 - (1/2)\hbar\omega + (\hbar\omega)^2/16U_0$$

Since $\omega = (K/\mu)^{1/2}$, we substitute $K = 573$ N/m (= 3.58 keV/nm^2), and for the reduced mass μ of H_2 one-half the proton mass to get

$$\hbar\omega = (197.3 \text{ eV}\cdot\text{nm/c})[2(3.58 \text{ keV/nm}^2)/(938.28 \times 10^3 \text{ keV/c}^2)]^{1/2}$$

$$= 0.545 \text{ eV}$$

With $E_{diss} = 4.52$ eV, the equation for U_0 becomes

$$4.52 = U_0 - 0.272 + 0.0186/U_0 \quad \text{or} \quad U_0 = 4.79 - 0.0186/U_0$$

A single iteration gives

$$U_0 \approx 4.79 - 0.0186/4.79 = 4.786 \text{ eV}$$

so that $U_0 \approx 4.79$ eV to the accuracy which can be expected from the given data. Then

$$\alpha = (K/2U_0)^{1/2} = \{(3.58 \times 10^3 \text{ eV/nm}^2)/[2(4.79 \text{ eV})]\}^{1/2}$$

$$= \underline{19.3} \text{ nm}^{-1}$$

---

14. The Morse levels are given by

$$E_{vib} = -U_0 + (\upsilon + 1/2)\hbar\omega - (\upsilon + 1/2)^2(\hbar\omega)^2/4U_0$$

The excitation energy from level υ to level $\upsilon + 1$ is

$$\Delta E_{vib} = (\upsilon + 3/2)\hbar\omega - (\upsilon + 3/2)^2(\hbar\omega)^2/4U_0 - (\upsilon + 1/2)\hbar\omega$$

$$- (\upsilon + 1/2)^2(\hbar\omega)^2/4U_0$$

$$= \hbar\omega - \{(\upsilon + 3/2)^2 - (v + 1/2)^2\}(\hbar\omega)^2/4U_0$$

$$= \hbar\omega\{1 - (\upsilon + 1)(\hbar\omega/2U_0)\}$$

It is clear from this expression that ΔE_{vib} diminishes steadily as υ increases. The excitation energy could never be negative, however, so that υ must not exceed the value which makes ΔE_{vib} vanish:

$$1 = (\hbar\omega/2U_0)(\upsilon + 1) \qquad \text{or} \qquad v_{max} = 2U_0/\hbar\omega - 1$$

With this value for υ, the vibrational energy is

$$E_{vib} = -U_0 + 2U_0 - (1/2)\hbar\omega - [2U_0 - (1/2)\hbar\omega]^2/4U_0$$

$$= -(\hbar\omega)^2/16U_0$$

If $2U_0/\hbar\omega$ is not an integer, then υ_{max} and the corresponding E_{vib} will be somewhat smaller than the values given. However, the maximum vibrational energy will never exceed $-(\hbar\omega)^2/16U_0$.

---

15. To the left and right of the barrier site ψ is the waveform of a free particle with wavenumber $k = (2mE/\hbar^2)^{1/2}$:

$$\psi(x) = A\sin(kx) + B\cos(kx) \qquad 0 \leq x \leq L/2$$

$$\psi(x) = F\sin(kx) + G\cos(kx) \qquad L/2 \leq x \leq L$$

The infinite walls at the edges of the well require $\psi(0) = \psi(L) = 0$, or $B = 0$ and $G = -F\tan(kL)$, leaving

$$\psi(x) = A\sin(kx) \qquad 0 \leq x \leq L/2$$

$$\psi(x) = F\{\sin(kx) - \tan(kL)\cos(kx)\} = C\sin(kx - kL) \qquad L/2 \leq x \leq L$$

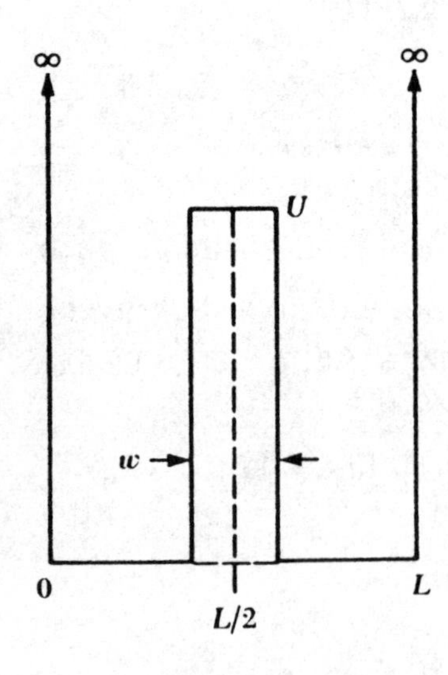

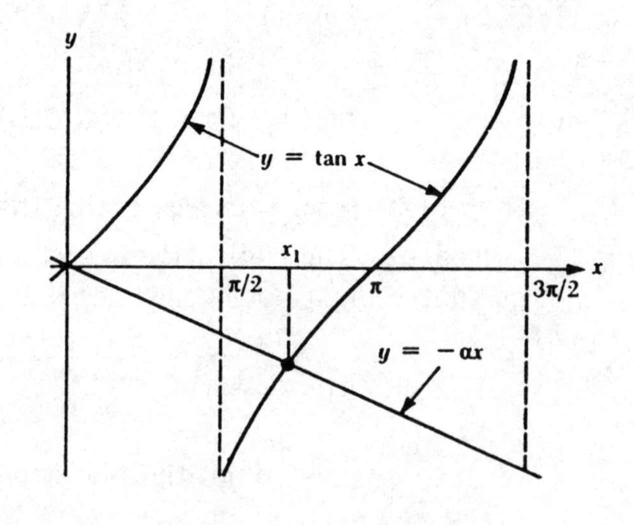

For waves antisymmetric about the midpoint of the well, $\psi(L/2) = 0$ and the delta barrier is ineffective: the slope $d\psi/dx$ is continuous at L/2, leading to C = +A. For this case $kL/2 = n\pi$, and

$$E_n = n^2\pi^2\hbar^2/2m(L/2)^2 , \qquad n = 1, 2, \ldots$$

as befits an infinite well of width L/2.

The remaining stationary states are waves symmetric about L/2, and require C = −A for continuity of ψ: their energies are found by applying the slope condition with C = −A to get

$$-Ak\cos(kL/2) - Ak\cos(kL/2) = (2mS/\hbar^2)A\sin(kL/2)$$

or

$$\tan(kL/2) = -(2\hbar^2/mSL)(kL/2)$$

Solutions to this equation may be found graphically as the intersections of the curve $y = \tan(x)$ with the line $y = -\alpha x$ having slope $-\alpha = -2\hbar^2/mSL$ (see Figure). From the points of intersection x_n we find $k_n = 2x_n/L$ and $E_n = \hbar^2 k_n^2/2m$. Only values of x_n greater than zero need be considered, since the wavefunction is unchanged when k is replaced by −k, and k = 0 leads to $\psi(x) = 0$ everywhere. As $S \to \infty$ we see that $x_n \to n\pi$, giving

$$E_n = n^2\pi^2\hbar^2/2m(L/2)^2 \qquad \text{for } S \to \infty \text{ and } n = 1,2, \ldots$$

the same energies found for the antisymmetric waves considered previously. Thus, in this limit the energy levels all are <u>doubly degenerate</u>.

As $S \to 0$ the roots become $x_n = \pi/2, 3\pi/2, \ldots = n\pi/2$ (n odd), giving

$$E_n = n^2\pi^2\hbar^2/2mL^2 \qquad n = 1, 3, \ldots$$

These are the energies for the symmetric waves of the infinite well with no barrier, as expected for $S = 0$.

The ground state wave is symmetric about $L/2$, and is described by the root x_1, which varies anywhere between $\pi/2$ and π according to S. The ground state energy is

$$E_1 = \hbar^2(2x_1/L)^2/2m = 2x_1{}^2\hbar^2/mL^2$$

The first excited state wave is antisymmetric, with energy

$$E_2 = \pi^2\hbar^2/2m(L/2)^2 = 2\pi^2\hbar^2/mL^2$$

which coincides with E_1 in the limit $S \to \infty$.

---

16. By trial and error, we discover that the choice $R = 2.49$ minimizes the expression for E_{tot}, so that this is the equilibrium separation R_0. The effective spring constant K is the curvature of $E_{tot}(R)$ evaluated at the equilibrium point $R_0 = 2.49$. Using the given approximation to

the second derivative with an increment $\Delta R = 0.01$, we find

$$K = d^2E_{tot}/dR^2\Big|_{R_0} \approx 0.126$$

(An increment ten times as large changes the result by only one unit in the last decimal place.) This value for K is in (Ryd/Bohr^2). The conversion to SI units is accomplished with the help of the relations

1 Ryd = 13.6 eV = 2.176×10^{-18} J, and 1 Bohr = 0.529 Å = 5.29×10^{-11} m.

Then

$$K = 0.126 \text{ Ryd/Bohr}^2 = 98.0 \text{ J/m}^2 = 98.0 \text{ N/m}$$

This is considerably smaller than the values reported for the hetero- nuclear diatomic molecules in Table 11.2. The latter form bonds with varying degrees of ionicity, and so tend to be more rigid structures.

---

17. By trial and error, we discover that the choice R = 1.44 bohrs minimizes the expression for E_{tot}, so that this is the equilibrium separation R_0.

The effective spring constant K is the curvature of $E_{tot}(R)$ evaluated at the equilibrium point R_0 = 1.44 bohrs. Using the given approximation to the second derivative with an increment ΔR = 0.01, we find

$$K = d^2E_{tot}/dR^2\Big|_{R_0} \approx 1.03$$

(An increment ten times as large changes the result by less than one unit in the last decimal place.) This value for K is in (Ryd/Bohr^2). The conversion to SI units is accomplished with the help of the relations

1 Ryd = 13.6 eV = 2.176×10^{-18} J, and 1 Bohr = 0.529 Å = 5.29×10^{-11} m

Then

$$K = 1.03 \text{ Ryd/Bohr}^2 = 801 \text{ J/m}^2 = 801 \text{ N/m}$$

The result is larger than the experimental value because our neglect of electron-electron repulsion leads to a potential well much deeper than the actual one, producing a larger curvature.

Chapter 12

1. $U_{Total} = U_{attractive} + U_{repulsive} = -\alpha k e^2/r + B/r^m$

At equilibrium, U_{TOTAL} reaches its minimum value.

$$dU_{Total}/dr = 0 = +\alpha k e^2/r^2 - m B/r^{m+1}$$

Calling the equilibrium separation r_0, we may solve for B

$$(m B)/r_0^{m+1} = \alpha k e^2/r_0^2$$

$$B = \alpha k e^2/m r_0^{m-1}$$

Substituting into the expression for U_{Total} we find

$$U_0 = -\alpha k e^2/r_0 + (\alpha k e^2/m) r_0^{m-1}/r_0^m = -(\alpha k e^2/r_0)(1 - 1/m)$$

---

3. $U = -\alpha k(e^2/r_0)(1 - 1/m)$

$U = -(1.7476)(9 \times 10^9)[(1.6 \times 10^{-19})^2/(0.281 \times 10^{-9})](1 - 1/8)$

$U = -1.25 \times 10^{-18}$ J = $\underline{-7.84}$ eV

---

4. (a) 199 kcal/mole

$= (199 \text{ kcal} \times 4186 \text{ J/kcal})/(6.02 \times 10^{23}\ Na^+ - Cl^- \text{ pairs})$

$= 1.384 \times 10^{-18}\ J/Na^+ Cl^-$

As $\quad |U_0| = (\alpha k e^2/r_0)(1 - 1/m)$,

$$r_0|U_0|/\alpha k e^2 = (1 - 1/m)$$

$$-r_0|U_0|/\alpha k e^2 + 1 = +1/m$$

or $\quad m = (1 - r_0|U_0|/\alpha k e^2)^{-1}$

4. (Cont'd) so

$$m = \left[1 - \frac{(0.257 \times 10^{-9})(1.384 \times 10^{-18})}{(1.7476)(9 \times 10^{9})(1.6 \times 10^{-19})^2}\right]^{-1} = 8.6 \approx 9$$

(b) $U_o = (\alpha k e^2/r_o)(7/8)$ for $m = 8$ $U_o = (\alpha k e^2/r_o)(8/9)$ for $m = 9$

The % change in $U_o = [(8/9 - 7/8)/7/8] \times 100 = 1.6\%$

---

5. $U = -ke^2/r - ke^2/r + ke^2/2r + ke^2/2r - ke^2/3r - ke^2/3r$

$+ ke^2/4r + ke^2/4r - \ldots$

$U = -2k(e^2/r)[1 - (1/2) + (1/3) - (1/4) + \ldots]$

but $\ln(1 + x) = x - (x^2/2) + (x^3/3) - (x^4/4) + \ldots$

so $U = -(2\ln 2)ke^2/r$

---

6. (a) $E_g = 1.14$ eV for Si

$hf = 1.14 \text{ eV} = (1.14 \text{ eV})(1.6 \times 10^{-19} \text{ J/eV}) = 1.82 \times 10^{-19}$ J

$f = \underline{2.75\ \ 10^{14}}$ Hz

(b) $c = \lambda f$; $\lambda = c/f = (3 \times 10^8 \text{ m/s})/(2.75 \times 10^{14} \text{ Hz}) = 1.09 \times 10^{-6}$ m

$\lambda = \underline{1090}$ nm (in the infrared region)

---

8. (a) $U_{Total} = -(k\alpha e^2/r) + B/r^m$

From Problem 1, note that $B = (k\alpha e^2/m)r_0^{m-1}$ which came from

using the condition $dU/dr\big]_{r = r_0} = 0$; Therefore,

$U_{Total} = -(k\alpha e^2/r) + (k\alpha e^2/m)(r_0^{m-1}/r^m)$

8. (Cont'd)

$$F = -\, dU_{Total}/dr = -\,(k\alpha e^2/r) + (k\alpha e^2/m)(-m)(r_0{}^{m-1}/r^{m-1})$$

$$= -\,(k\alpha e^2/r^2)\left[\,1 - (r_0/r)^{m-1}\right]$$

(b) If the atom is displaced by a <u>small</u> distance x from r_0, we can replace r with $r_0 + x$, and F becomes

$$F = -\,[k\alpha e^2/(r_0 + x)^2][1 - r_0{}^{m-1}/(r_0 + x)^{m-1}] \quad \text{Note: } (x/r_0 << 1)$$

$$= -\,[k\alpha e^2/r_0{}^2(1 + x/r_0)^2][1 - 1/(1 + x/r_0)^{m-1}]$$

$$\approx -\,(k\alpha e^2/r_0{}^2)(1 - 2x/r_0)\{1 - 1/[1 + (m-1)(x/r_0) + \ldots]\}$$

$$\approx -\,(k\alpha e^2/r_0{}^2)(1 - 2x/r_0)[1 + (m-1)(x/r_0) - 1]/[1 + (m-1)(x/r_0) + .\,]$$

$$F \approx -\,(k\alpha e^2/r_0{}^2)(1 - 2x/r_0)(m-1)x/r_0 \approx -\,(k\alpha e^2/r_0{}^3)(m-1)x$$

(Neglecting the term in x/r_0 in the denominator). Therefore, the ion experiences a <u>restoring</u> force $F = -\,Kx$, where the effective force constant K is

$$K \approx (k\alpha e^2/r_0{}^3)(m-1) \quad \text{Q.E.D.}$$

(c) For NaCℓ, $\alpha = 1.7476$, $r_0 = 0.281\text{ nm} = 0.281 \times 10^{-9}\text{ m}$, $m = 8$ so that

$$K \approx (9 \times 10^9)(1.7476)(1.6 \times 10^{-19})^2(7)/(0.281 \times 10^{-9})^3 \approx 127\text{ N/m}$$

The corresponding frequency of vibration is

$$f = (1/2\pi)(K/m)^{1/2}$$

where we use the mass of Na to estimate f; $m = 23 \times 1.67 \times 10^{-27}$ kg, so

$$f = (1/2\pi)[(126\text{ N/m})/(23 \times 1.67 \times 10^{-27})]^{1/2} = \underline{9.15 \times 10^{12}}\text{ Hz}$$

10. (a) $\int_0^{\infty} (Ne/\tau)\, e^{-t/\tau}\, dt = -\, Ne^{-t/\tau}\Big|_0^{\infty} = -N\,[e^{-\infty} - e^0] = N$

(b) $\langle t \rangle = (1/N)\int_0^{\infty} (tN/\tau)\, e^{-t/\tau}\, dt = \tau \int_0^{\infty} (t/\tau)\, e^{-t/\tau}\, dt/\tau$

$= \tau \int_0^{\infty} ze^{-z}\, dz \qquad z = u \qquad dv = e^{-z}\, dz$

$\qquad dz = du \qquad v = -e^{-z}$

so $\int_0^{\infty} ze^{-z}\, dz = (-ze^{-z})\Big|^{\infty} + \int_0^{\infty} e^{-z}\, dz = 0 - e^{-z}\Big|_0^{\infty} = 1$

Therefore, $\langle t \rangle = \tau$

c) Similarly, $\langle t^2 \rangle = (1/N) \int_0^{\infty} (t^2\, N/\tau)\, e^{-t/\tau}\, dt$

Integrating by parts twice, gives $\qquad \langle t^2 \rangle = 2\tau$

---

11. a) $\tau = m_e/ne^2\rho$

n = number of electrons/m^3

$= (1 \text{ electron/atom})(6.02 \times 10^{26} \text{ atoms/k mole})$

$\times (10.5 \times 10^3 \text{ kg/m}^3)(1 \text{ kmole/108 kg})$

$= 5.85 \times 10^{28}$ electrons/m^3

11. (Cont'd) so

$$\tau = \frac{(9.11 \times 10^{-31}\ kg)}{(1.60 \times 10^{-8}\ \Omega\cdot m)(5.85 \times 10^{28}\ m^{-3})(1.60 \times 10^{-19}\ C)^2}$$

$= \underline{3.80 \times 10^{-14}\ s}$

(b) As shown in example 12.1, the rms speed of an electron at room temperature is about 1.2×10^5 m/s. In 3.80×10^{-14} s the electron would travel a distance

$$L = v_{rms}\cdot\tau = (1.2 \times 10^5\ m/s) \times 3.80 \times 10^{-14}\ s$$

$= 4.6 \times 10^{-9}$ m $= \underline{46\ Å}$ or 10 - 20 lattice spacings.

---

12. (a) There are 6 Cl^- ions at r_0 : $U_c = -6ke^2/r_0$

There are 12 Na^+ ions at $\sqrt{2}\,r_0$: $U_c = +12ke^2/\sqrt{2}\,r_0$

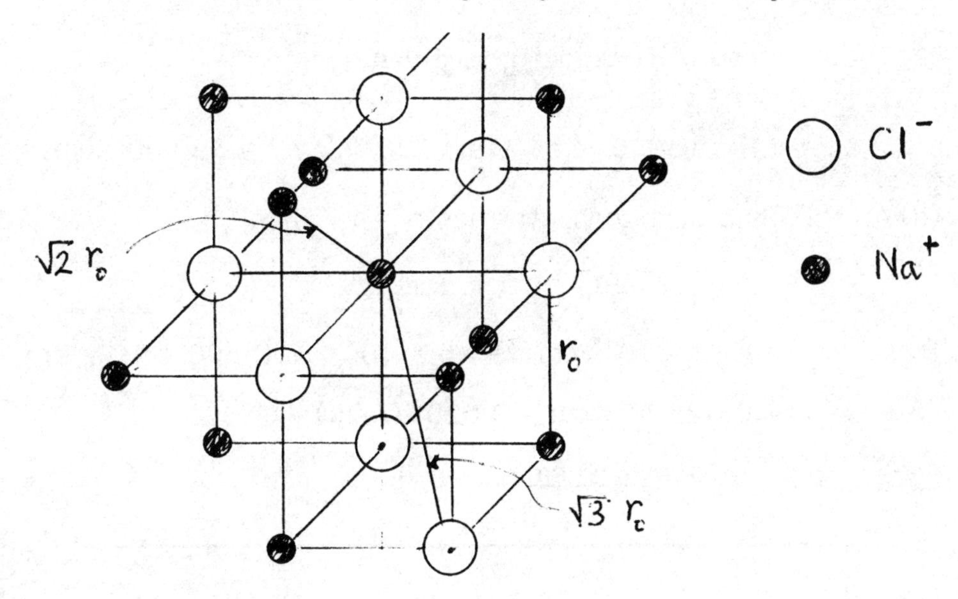

There are 8 Cl^- ions at $\sqrt{3}\,r_0$: $U_c = -8ke^2/(\sqrt{3}\,r_0)$

So U_C to three terms is:

$$U_c = [-6 + 12/\sqrt{2} - 8/\sqrt{3}]ke^2/r_0 = -2.13ke^2/r_0$$

12. (b) The fourth term consists of 6 Na^+ ions at $2r_0$:

$$U_c = (-2.13 + 3)\ ke^2/r_o = 0.87\ ke^2/r_o$$

So we see that the Coulomb potential is not even attractive to 4 terms, and that the infinite series does not converge rapidly when groups of atoms corresponding to nearest neighbors, next-nearest neighbors, etc. are added together.

---

13. (a) Eq. 12.14 was $J = nev_d$. As $v_d = \mu E$, $J = ne\mu E$.

Also comparing Eq. 12.13, $v_d = e\tau E/m_e$, and $v_d = \mu E$, one has

$$\mu = e\tau/m_e$$

(b) As $J = \sigma E$ and $J = J_{electrons} + J_{holes} = ne\mu_n E + pe\mu_p E$,

$$\sigma = ne\mu_n + pe\mu_p$$

(c) The electron drift velocity is given by

$$v_d = \mu_n E = \left(3900\ \text{cm}^2/\text{V}\cdot\text{s}\right)\left(100\ \text{V/cm}\right) = 390{,}000\ \text{cm/s}$$

(d) An intrinsic semiconductor has $n = p$. Thus

$$\sigma = ne\mu_n + pe\mu_p = pe(\mu_n + \mu_p)$$

$$= (3.0 \times 10^{13}\ \text{cm}^{-3})(1.6 \times 10^{-19}\ \text{C})(5800\ \text{cm}^2/\text{V}\cdot\text{s})$$

$$= 0.028\ \text{A/V}\cdot\text{cm} = 0.028\ (\Omega\cdot\text{cm})^{-1}$$

$$\rho = 1/\sigma = \underline{36\ \Omega\cdot\text{cm}}$$

---

14. (a) $v_d = \mu E = (8500\ \text{cm}^2/\text{V}\cdot\text{s})\ (10\ \text{V/cm}) = \underline{85{,}000\ \text{cm/s.}}$

(b) $(1/2)mv^2_{th} = 3kT/2$

$$v_{th} = (3kT/m)^{1/2} = \left[(3)(1.40 \times 10^{-23}\ \text{J/K})(300\ \text{K})\ /\ 9.1 \times 10^{-31}\ \text{kg}\right]^{1/2}$$

$$= 117{,}000\ \text{m/s} = 1.17 \times 10^7\ \text{cm/s}$$

14. (Cont'd)

$$v_d/v_{th} \times 100 = (8.50 \times 10^4)/(1.17 \times 10^7) \times 100 = \underline{0.73\ \%}$$

(c)From Equation 12.13, $v_d/E = \mu = e\tau/m_e$ so

$$\tau = m_e\mu/e = (9.11 \times 10^{-31}\ \text{kg})(8500\ \text{cm}^2/\text{V}\cdot\text{s})/(1.60 \times 10^{-19}\ \text{C})$$

$$= 4.8 \times 10^{-8}\ \text{kg}\cdot\text{cm}^2/\text{J}\cdot\text{s}$$

$$= 4.8 \times 10^{-8}\ (\text{kg}\cdot\text{cm}^2)/[(\text{kg}\cdot\text{m}^2/\text{s}^2)\cdot\text{s}]$$

$$= 4.8 \times 10^{-12}\ \text{s}$$

(d)$L = v_{th}\cdot\tau = (1.17 \times 10^7\ \text{cm/s})(4.8 \times 10^{-12}\ \text{s}) = 5.6 \times 10^{-5}\ \text{cm}$

$= \underline{5600\ \text{Å}}$

---

15. (a)For Si, $\text{B. E.} = 13.6(Z/\kappa)^2\ \text{eV} = (13.6)(1/12)^2\ \text{eV} = 0.094\ \text{eV}$

For Ge, $\text{B. E.} = (13.6)(1/16)^2 = 0.053\ \text{eV}$

Since $kT = 0.03$ eV at room temperature, there is enough thermal energy to ionize many donor electrons and promote them to the conduction band.

(b)For Si, $r_1 = a_0/Z = a_0/(1/\kappa) = \kappa a_0 = 12a_0 = 6.4\ \text{Å}$

For Ge, $r_1 = \kappa a_0 = 16a_0 = 8.5\ \text{Å}$. Thus, the donor electron is not localized but roams over a radius of 3 or 4 semiconductor atoms.

---

16. (a)We assume all expressions still hold with $\langle v\rangle$ replaced by v_F.

$$\tau = \sigma m_e/ne^2$$

$$\sigma = 1/\rho = (1.60 \times 10^{-8})^{-1}(\Omega\cdot\text{m})^{-1} = 6.25 \times 10^{17}\ (\Omega\cdot\text{m})^{-1}$$

$$n = \#\text{ of } e^-/\text{m}^3 = (1\ e^-/\text{atom})(6.02 \times 10^{26}\ \text{atoms/k mole})$$

$$\times (10.5 \times 10^3\ \text{kg/m}^3)(1\ \text{kmole}/108\ \text{kg})$$

16. (Cont'd)

$$n = 5.85 \times 10^{28}\ e^-/m^3$$

So $\tau = \dfrac{(6.25 \times 10^7)(\Omega \cdot m)^{-1}(9.11 \times 10^{-31}\ kg)}{(5.85 \times 10^{28}\ e^-/m^3)(1.60 \times 10^{-19}\ C)^2}$

$= \underline{3.80 \times 10^{-14}\ s}$ (no change of course from 12.11)

(b) Now $L = v_F \cdot \tau$ and $v_F = (2E_F/m)^{1/2}$

$$v_F = \left[(2 \times 5.48\ eV \times 1.6 \times 10^{-19}\ J/eV)/(9.11 \times 10^{-31}\ kg)\right]^{1/2}$$

$= \underline{1.39 \times 10^6\ m/s}$.

$L = (1.39 \times 10^6\ m/s)(3.8 \times 10^{-14}\ s) = 5.27 \times 10^{-8}\ m$

$= \underline{52.7\ nm} = \underline{527\ Å}$

(c) The approximate lattice spacing in silver may be calculated from the density and the molar weight. The calculation is the same as the n calculation. Thus,

$$(\#\ \text{of Ag atoms})/m^3 = 5.85 \times 10^{28}$$

Assuming each silver atom fits in a cube of side, d,

$$d^3 = (5.85 \times 10^{28})^{-1}\ m^3/\text{atom}$$

$$d = \underline{2.57 \times 10^{-10}\ m}$$

So $L/d = (5.27 \times 10^{-8})/(2.57 \times 10^{-10}) = 205$

---

17. (a) $n = \#$ of electrons/cm^3

$n = (1e^-/\text{atom})(6.02 \times 10^{23}\ \text{atoms/mole})(0.971\ g/cm^3)(1\ \text{mole}/23.0\ g)$

$= \underline{2.54 \times 10^{22}\ e/cm^3}$

(b) $E_F = (h^2/2m)(3n/8\pi)^{2/3}$

$$= \frac{(6.625 \times 10^{-34}\ J\cdot s)^2\ (3 \times 2.54 \times 10^{28}\ m^{-3})}{2(9.11 \times 10^{31}\ kg)(8\pi)}$$

17. (Cont'd)

$E_F = (2.409 \times 10^{-37})(2.09 \times 10^{18}) = 5.04 \times 10^{-19}\ \text{J} = \underline{3.15\ \text{eV}}$

(c) $v_F = \left(2E_F/m_e\right)^{1/2} = \left[2(5.04 \times 10^{-9}\ \text{J})/(9.11 \times 10^{-31}\ \text{kg})\right]^{1/2}$

$= 1.05 \times 10^6\ \text{m/s}$

(d) $\tau = m_e/\rho n e^2$

$= (9.11 \times 10^{-34}\ \text{kg})/(4.20 \times 10^{-8}\ \Omega\cdot\text{m})(2.54 \times 10^{28}\ e^-/\text{m}^3)(1.6 \times 10^{-19}\ \text{C})$

or $\tau = \underline{3.33 \times 10^{-14}}\ \text{s}$

(e) $L = v_F \cdot \tau = (1.05 \times 10^6)(3.33 \times 10^{-14}\ \text{s}) = 35.0 \times 10^{-9}\ \text{m} = \underline{350\ \text{Å}}$

L/nearest neighbor distance = 94

(f) $K/\sigma T = \pi^2 k^2/3e^2 = 2.45 \times 10^{-8}\ \text{W}\cdot\Omega/\text{K}^2$ or

$K = \left(2.45 \times 10^{-8}\ \text{W}\cdot\Omega/\text{K}^2\right)\left(2.38 \times 10^7\ (\Omega\cdot\text{m})^{-1}\right)\left(300\ \text{K}\right) = \underline{175\ \text{W/m}\cdot\text{K}}$

Chapter 13

1. $R = R_0 A^{1/3}$ where $R_0 = 1.2$ fm ;

   a) $A = 4$ so $R_{He} = (1.2)(4)^{1/3}$ fm = 1.9 fm

   b) $A = 238$ so $R_U = (1.2)(238)^{1/3}$ fm = 7.44 fm

   c) $R_U/R_{He} = (7.44 \text{ fm})/(1.9 \text{ fm}) = 3.92$

---

2. $m = \rho V = (2.3 \times 10^{14} \text{g/cm}^3)(10 \text{cm}^3) = 2.3 \times 10^{15}$g ! ! !

---

3. $\rho_{NUC}/\rho_{ATOMIC} = (M_{NUC}/V_{NUC})/(M_{ATOMIC}/V_{ATOMIC})$

   and approximately; $M_{NUC} = M_{ATOMIC}$

   Therefore $\rho_{NUC}/\rho_{ATOMIC} = (r_0/R)^3$

   where $r_0 = 0.529$ Å $= 5.29 \times 10^{-11}$ m (Eq. 13.1)

   and $R = 1.2 \times 10^{-15}$ m (Eq. 13.1 where A = 1)

   So that

   $\rho_{NUC}/\rho_{ATOMIC} = [(5.29 \times 10^{-11} \text{ m})/(1.2 \times 10^{-15} \text{ m})]^3 = 8.57 \times 10^{13}$

---

4. (a) $f_n = |2mB|/h$

   $f_n = 2(1.9135)(5.05 \times 10^{-27} \text{ J/T})(1 \text{ T})/(6.63 \times 10^{-34} \text{ J·s}) = 29.1$ MHz

(b) $f_p = 2(2.7928)(5.05 \times 10^{-27} \text{ J/T})(1 \text{ T})/6.63 \times 10^{-34} \text{ J·s}) = 42.5$ MHz

(c) In the earth's magnetic field

   $f_p = 2(2.7928)(5.05 \times 10^{-27})(50 \times 10^{-6})/(6.63 \times 10^{-34}) = 2.13$ KHz

---

5. (a) The initial kinetic energy of the alpha particle must equal the electrostatic potential energy of the two particle system at the distance of closest approach ; $K_\alpha = U = kqQ/r_{min}$ and

$r_{min} = kqQ/K_\alpha$

$r_{min} = (9 \times 10^9)2(79)(1.6 \times 10^{-19})^2/[0.5(1.6 \times 10^{-13})] = \underline{4.55 \times 10^{-13}}$ m

(b) Note that $K_\alpha = 1/2mv^2 = kqQ/r_{min}$, so

$v = [2kqQ/(mr_{min})]^{1/2}$

$v = \{2(9 \times 10^9)2(79)(1.6 \times 10^{-19})^2/[4(1.67 \times 10^{-27})(3 \times 10^{-13})]\}^{1/2}$

$v = \underline{6.03 \times 10^6}$ m/s

---

6. $E_b/A = (1/3)[1(1.007276) + 2(1.008665) - 3.01605]\ (931.5)$

$= \underline{2.657}$ MeV/nucleon

---

7. $E_b = [Zm_{1H} + Nm_n - m(45Fe)](931.5\ MeV/u)$

$E_b = \underline{492}$ MeV

$E_b/A = 492/56 = \underline{8.79}$ MeV/nucleon ; agrees with Fig. 13.7 .

---

8. (a) We make the calculation for $r = R = 3$ fm to get;

$E = ke^2/r = (9 \times 10^9\ N{\cdot}m^2/C^2)(1.6 \times 10^{-19}\ C)^2/(3 \times 10^{-15}\ m)$

$E = 7.68 \times 10^{-14}\ J = \underline{0.480}$ MeV

(b) For the electron;

$m_0c^2 = (9.11 \times 10^{-31}\ kg)(3 \times 10^8\ m/s)^2 = 8.2 \times 10^{-14}\ J = \underline{0.512}$ MeV

---

9. For $^{15}_{8}O$ we have, using Eq. 13.4

$$E_b = [(8)(1.007825) + 7(1.008665) - (15.003065)]\ u\ (931.5\ MeV/u)$$

$$E_b = 111.96\ MeV$$

For $^{15}_{7}N$ we have

$$E_b = [(7)(1.007825) + 8(1.008665) - (15.000109)]\ u\ (931.5\ MeV/u)$$

$E_b = 115.49$ MeV Therefore $\Delta E_b = \underline{3.53}$ MeV

---

10. Use Equation 13.4

(a) For $^{20}_{10}Ne$; $E_b/A = \underline{8.03}$ MeV/nucleon

(b) For $^{40}_{20}Ca$; $Eb/A = \underline{8.55}$ MeV/nucleon

(c) For $^{93}_{41}N_b$; $E_b/A = \underline{8.66}$ MeV/nucleon

(d) For $^{197}_{79}Au$; $E_b/A = \underline{7.92}$ MeV/nucleon

---

11. Removal of a neutron from $^{43}_{20}Ca$ would result in the residual nucleus, $^{42}_{20}Ca$. If the required separation energy is S_n, the overall process can be described by

$$\text{mass}(^{43}_{20}Ca) + S_n = \text{mass}(^{42}_{20}Ca) + \text{mass}(n) \quad \text{or}$$

$$S_n = (41.958625 + 1.008665 - 42.98780)\ u$$

$$S_n = (0.008510\ u)(931.5\ MeV/u) = \underline{7.93}\ MeV$$

---

12. Use Equation 13.4 to find

for $^{23}{}_{11}Na$; $E_b/A = 8.11$ MeV/nucleon

and for $^{23}{}_{12}Mg$; $E_b/A = 7.90$ MeV/nucleon

The binding energy per nucleon is greater for $^{23}{}_{11}Na$ by 0.210 MeV

---

13. (a) The neutron to proton ratio , $(A - Z)/Z$ is greatest for $^{139}{}_{55}C_s$ and is equal to 1.53.

(b) $^{139}L_a$ has the largest binding energy per nucleon of 8.379 MeV

(c) $^{139}C_s$ with a mass of 138.913 u

---

14. (a) The first term overstates the importance of volume and the second term subtracts this overstatement.

(b) For spherical volume $(4/3)\pi R^3/(4\pi R^2) = R/3$

For cubical volume $R^3/(6R^2) = R/6$

The maximum binding energy or lowest state of energy is achieved by building "nearly" spherical nuclei.

---

15. $\Delta E = E_{bf} - E_{bi}$

For $A = 200$; $E_b/A = 7.8$ MeV so

$$E_{bi} = (A_i)(7.8 \text{ MeV}) = (200)(7.8) = 1560 \text{ MeV}$$

For $A \approx 100$; $E_b/A \approx 8.6$ MeV so

$$E_{bf} = (2)(100)(8.6 \text{ MeV}) = (200)(8.6) = 1720 \text{ MeV}$$

$\Delta E = E_{bf} - E_{bi} = 1720 \text{ MeV} - 1560 \text{ MeV} = 160$ MeV

---

16. (a) For ^{64}Cu, $E_b = (15.7\text{ MeV})(64) - (17.8\text{ MeV})(64)^{2/3}$

$- (0.71\text{ MeV})[(29)(28)/(64)^{1/3}] - (23.6\text{ MeV})[64 - 2(29)]^2/64$

$E_b = 562.6\text{ MeV}$ or $\underline{8.79}$ MeV/nucleon

For ^{64}Zn a similar calculation gives $\underline{8.75}$ MeV/nucleon

(b) For $^{64}{}_{29}Cu$,

$E_b = [(29)(1.007825) + (93.4)(1.00865) - 63.929766](931.5)\text{ MeV}$

$E_b = 559.3\text{ MeV}$ or $\underline{8.74}$ MeV/nucleon

For ^{64}Zn, $E_b = [(30)(m_H) + (34)(m_N) - 63.929145](931.5)\text{ MeV}$

$E_b = 559.3\text{ MeV}$ or $\underline{8.74}$ MeV/nucleon

---

17. (a) Write Eq. 13.8 as $R/R_0 = e^{-\lambda t}$ so that $\lambda = (1/t)\ln(R_0/R_1)$

In this case $R_0/R_1 = 5$ when $t = 2$ h, so

$\lambda = (1/2\text{ h})\ln 5 = \underline{0.805}\text{ h}^{-1}$

(b) From Eq. 13.9, $T_{1/2} = (\ln 2)/\lambda = (\ln 2)/(0.805\text{ h}^{-1}) = \underline{0.861}$ h

---

18. (a) $\lambda = \ln 2/T_{1/2} = 0.693/8.04\text{ day} = \underline{9.98 \times 10}^{-7}\text{ s}^{-1}$

(b) $|dN/dt| = R = \lambda N$ so $N = R/\lambda$

$N = (0.5 \times 10^{-6})(3.7 \times 10^{10})/(9.98 \times 10^{-7}) = \underline{1.85 \times 10}^{10}$ nuclei

---

19. (a) From $R = R_0 e^{-\lambda t}$, $\lambda = (1/t)\ln(R_0/R)$

$\lambda = (1/4\text{ hr})\ln(10/8) = 5.58 \times 10^{-2}\text{ hr}^{-1} = \underline{1.55 \times 10}^{-5}\text{ s}^{-1}$

and $T_{1/2} = \ln 2/\lambda = \underline{12.4}$ hr

(b) $R_0 = 10\text{ mCi} = 10 \times 10^{-3} \times 3.7 \times 10^{10}$ decays/s $= 3.7 \times 10^8$ decays/s and $R = \lambda N$ so

$N_0 = R_0/\lambda = 3.7 \times 10^8/(1.55 \times 10^{-5}) = \underline{2.39 \times 10}^{13}$ atoms

19. (c) $R = R_0 e^{-\lambda t} = (10\ \text{mCi})e^{-(5.58 \times 10^{-2})(30)} = \underline{1.87}\ \text{mCi}$

---

20. From Eq. 13.7, the fraction of nuclei remaining after five years will be

$$N/N_0 = e^{-\lambda t} = e^{-(\ln 2/T_{1/2})t} = e^{-(0.693/12.33)(5)} = 0.755$$

Therefore the percentage decaying during the interval will be

$(1 - 0.755)$ or $\underline{24.5}\%$

---

21. Combining Eqs. 13.6 and 13.9 we have

$$N = |dN/dt|/\lambda = |dN/dt|/(0.693/T_{1/2})$$

and since 1 mCi = 3.7×10^7 decays/s

$$N = (5\ \text{mCi})(3.7 \times 10^7\ \text{dps/mCi})/\{(0.693)/[(28.8\ \text{yr})(3.16 \times 10^7\ \text{s/yr})]\}$$

$$N = 2.43 \times 10^{17}\ \text{atoms}$$

Therefore, the mass of strontium in the sample is

$$m = (N/N_A)M = [(2.43 \times 10^{17})/(6.022 \times 10^{23})](90) = \underline{36.3}\ \mu\text{g}$$

---

22. $dN/dt = -\lambda N$ and $T_{1/2} = 0.693/\lambda$ thus

$|dN/dt| = R = \lambda N = 0.693\, N/T_{1/2}$ and $1\ \mu\text{Ci} = 3.7 \times 10^4$ decays/s

$N = R/(0.693/T_{1/2})$

$N = (0.2)(3.7 \times 10^4)/0.693/(8.1 \times 8.64 \times 10^4) = \underline{7.47 \times 10^9}$ atoms

---

23. Let R_0 equal the total activity withdrawn from the stock solution.

$$R_0 = (2.5\ \text{mCI/ml})(10\ \text{ml}) = 25\ \text{mCi}.$$

Let R_0' equal the initial specific activity of the working solution.

$$R_0' = 25\ \text{mCi}/250\ \text{ml} = 0.1\ \text{mCi/ml}$$

After 48 hours the specific activity of the working solution will be

$$R' = R_0' e^{-\lambda t} = (0.1\ \text{mCi/ml})e^{-(0.693/15\ \text{h})(48\ \text{h})} = 0.011\ \text{mCi/ml}$$

and the activity in the sample will be,

$R = (0.011\ \text{mCi/ml})(5\ \text{ml}) = \underline{0.055}\ \text{mCi}$

---

24. (a) From $R = R_0 e^{-\lambda t}$, $\ell n(R/R_0) = -\lambda t$

and $\lambda = (1/t)\ell n(R_0/R)$

(b) $T_{1/2} = \ell n2/\lambda = \ell n2/[(1/t)\ell n(R_0/R)] = (\ell n2)(t)/\ell n(R_0/R)$

---

25. The number of nuclei which decay during the interval will be

$N_1 - N_2 = N_0(e^{-\lambda t_1} - e^{-\lambda t_2})$

First we find λ;

$\lambda = \ln 2/T_{1/2} = 0.693/64.8\ \text{h} = 0.0107\ \text{h}^{-1} = 2.97 \times 10^{-6}\ \text{s}^{-1}$ and

$N_0 = R_0/\lambda = (40\ \mu\text{Ci})(3.7 \times 10^4\ \text{dps}/\mu\text{Ci})/(2.97 \times 10^{-6}\ \text{s}^{-1})$

$N_0 = 4.98 \times 10^{11}$ nuclei

Using these values we find

$$N_1 - N_2 = (4.98 \times 10^{11})\left[e^{-(0.0107\ \text{h}^{-1})(10\ \text{h})} - e^{-(0.0107\ \text{h}^{-1})(12\ \text{h})}\right]$$

Hence, the number of nuclei decaying during the interval is

$N_1 - N_2 = \underline{9.46 \times 10^9}$ nuclei

---

26. $Q = (M_{^{238}U} - M_{^{234}Th} - M_{^4He})(931.5)$

$= (238.048608 - 234.043583 - 4.002603)(931.5) = \underline{2.26}\ \text{MeV}$

---

27. (a) $\Delta = (M_N - 1) = (1.008665 - 1) = \underline{0.008665}\ \mu$

or $\Delta = (8.665 \times 10^{-3})(931.5) = \underline{8.07}\ \text{MeV}$

(b) $\Delta = (M_{^{63}Cu} - 1) = (62.929599 - 63) = -\underline{0.0704}\ \mu = -\ \underline{65.68}\ \text{MeV}$

---

28. $Q = [m(R_n) - m(He) - m(P_0)]931.5$ MeV

$Q = [220.011401 - 4.002603 - 216.001790]931.5 = \underline{6.53}$ MeV

If we assume no recoil, $K_\alpha = Q = \underline{6.53}$ MeV

---

29. (a) $Q = (m_{initial} - m_{final})931.5$

$= (39.96259 - 0.0005486 - 39.96459)931.5 = -\ \underline{2.37}$ MeV

$Q < 0$ so the reaction cannot occur

(b) $Q = (97.9055 - 4.0026 - 93.9047(931.5 = -\ \underline{1.68}$ MeV

$Q < 0$ so the reaction cannot occur

(c) $Q = (143.9099 - 4.0026 - 139.9053)931.5 = 1.86$ MeV

$Q > 0$ so the reaction can occur

30.

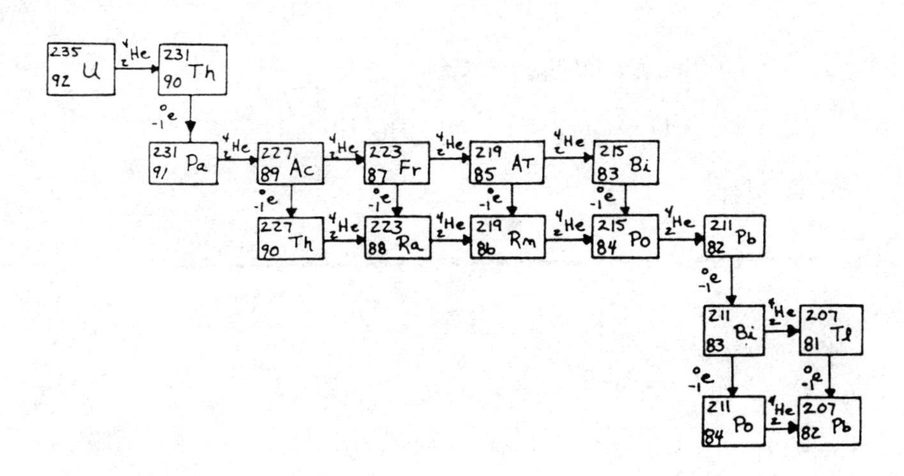

---

31. $Q = (M_\alpha + M_{(^9Be)} - M_{(^{12}C)} - M_n)(931.5\ MeV/u)$

$= (4.002603 + 9.012182 - 12.00000 - 1.008665)(931.5)$

$Q = \underline{5.70}\ MeV$

---

32. $Q = [M_\alpha + M_{(^{27}Al)} - M_{(^{30}P)} - M_n](931.5)$

$= (4.002603 + 26.981541 - 29.978310 - 1.008665)(931.5)$

$Q = -\ \underline{2.64\ MeV}$

---

33. $Q = (m_a + m_X - m_Y - m_b)[931.5\ MeV/u]$

$Q = [m_P + m(^7Li) - m(^4He) - m_\alpha)[931.5\ MeV/u]$

$Q = [(1.007825 + 7.016004 - 4.002603 - 4.002603)\ u][931.5\ MeV/u]$

$Q = \underline{17.35}\ MeV$

---

34. $Q_1 = [M(^2{}_4He) + M(^{10}{}_5B) - M(^1{}_1H) - M(^{13}{}_6C)]\,c^2$

$Q_2 = [M(^{13}{}_6C) + M(^1{}_1H) - M(^{10}{}_5B) - M(^4{}_2He)]\,c^2$

$Q_1 = -\ Q_2\ ;\quad |Q_1| = |Q_2|$

---

35. (a) $^1{}_1H + {}^{18}{}_8O \rightarrow {}^1{}_0n + {}^{18}{}_9F$

(b) $Q = [M(^{18}{}_8O) + M(^1{}_1H) - M(^{18}{}_9F) - M(^1{}_0n)]\ c^2$

so $M(^{18}{}_9F) = [M(^{18}{}_8O) + M(^1{}_1H) - M(^1{}_0n)] - Q/c^2$

$M(^{18}{}_9F) = (18.000934 \pm 0.000002)\ u$

Tabulated value = 18.000936 u

---

36. $^{9}Be + n \rightarrow {}^{10}Be + 6.810$ MeV

$m(^{10}Be) = m(^{9}Be) + m_n - (6.810 \text{ MeV})/(931.5 \text{ MeV/u})$

$= 9.01218 + 1.008665 - 0.007311 = \underline{10.0135}$ u

$^{9}Be + 1.666 \text{ MeV} \rightarrow {}^{8}Be + n$

$m(^{8}Be) = m(^{9}Be) - m_n + (1.66 \text{ MeV })/(931.5 \text{ MeV/u})$

$= 9.01218 - 1.008665 + 0.001782 = \underline{8.0053}$ u

---

37. We need to use the procedure to calculate a "weighted average." Let the fractional abundances be represented by $f_{63} + f_{65} = 1$; then

$$[f_{63}\, m(^{63}Cu) + f_{65}\, m(^{65}Cu)]/(f_{63} + f_{65}) = m_{Cu}$$

We find $f_{63} = [m(^{65}Cu) - m_{Cu}]/[m(^{65}Cu) - m(^{63}Cu)]$

$f_{63} = (64.95 - 63.55)/(64.95 - 62.95) = \underline{0.30}$ or $\underline{30}\%$

and $f_{65} = 1 - f_{63} = \underline{0.70}$ or $\underline{70}\%$

---

38. (a) $Q = [m(^{14}N) + m(^{4}He) - m(^{17}O) - m(^{1}H)](931.5 \text{ MeV/u})$

$Q = (14.003074 + 4.002603 - 16.999131 - 1.007825)(931.5)$ MeV

$Q = -1.19$ MeV

$E_{th} = - Q[m(^{4}He) + m(^{14}N)]/m(^{14}N)]$

$= - (- 1.19 \text{ MeV})(4.002603/14.003074) = \underline{1.53}$ MeV

(b) $Q = [(m(^{7}Li) + m(^{1}H) - 2m(^{1}He)](931.5 \text{ MeV/u})$

$Q = [(7.016004 + 1.007825) - (2)(4.002603) \text{ u}](931.5 \text{ MeV/u})$

$Q = \underline{17.35}$ MeV

---

39. (a) $Q = [m(^9Be) + m(^4He) - m(^{12}C) - m(^1n)](931.5 \text{ MeV/u})$

$Q = [(9.01292 + 4.002603 - 12.0000 - 1.008665) \text{ u}](931.5 \text{ MeV/u})$

$Q = \underline{5.70} \text{ MeV}$

(b) $Q = [2m(^2H) - m(^3He) - m(^1n)](931.5 \text{ MeV/u})$

$Q = [(2)(2.014102) - (3.016029) - 1.008665 \text{ u}](931.4 \text{ MeV/u})$

$Q = \underline{3.27} \text{ MeV}$; <u>exoergic</u>

---

40. (a)

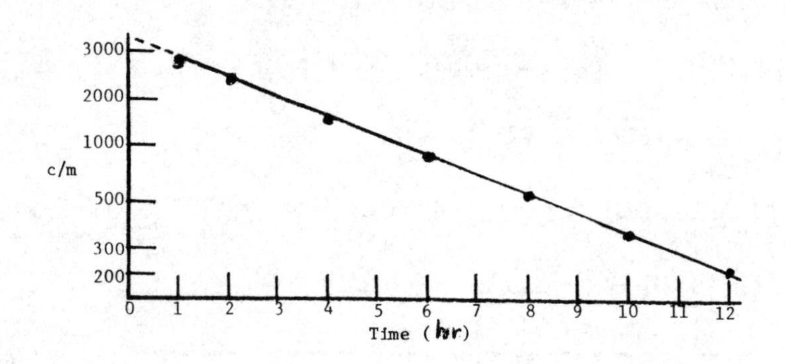

(b) $\lambda = -\text{slope} = -(\ln 200 - \ln 1480)/[(12 - 4) \text{ hr}]$

$= \underline{0.25 \text{ hr}^{-1}} = 4.17 \times 10^{-3} \text{ min}^{-1}$

and $T_{1/2} = (\ln 2)/\lambda = \underline{2.77} \text{ hr}$

(c) By extrapolation of graph to t = 0, we find $(\text{cpm})_0 = \underline{4 \times 10^3} \text{ cpm}$

(d) $N = R/\lambda$; $N_0 = R_0/\lambda = [(\text{cpm})_0/\text{EFF}]/\lambda$

$N_0 = (4 \times 10^4 \text{ dis/min})/(4.17 \times 10^{-3} \text{ min}^{-1}) = \underline{9.59 \times 10^6} \text{ atoms}$

---

41. (a) # nuclei = [(1000 g)/(94 g)](6.02×10^{23}) = $\underline{6.40 \times 10^{24}}$;

$$\lambda = 9.16 \times 10^{-13}$$

(b) $$R_0 = \lambda N_0$$

Taking the half life to be 24000 yrs = 7.65×10^{11} s, and using the expression $\lambda = (\ln 2)/T_{1/2}$, we find

$$R_0 = 58.6 \times 10^{11} \text{ decays/s}$$

(c) $R = R_0 e^{-\lambda t}$ or $\ln(R/R_0) = -\lambda t$ or $t = -[\ln(R/R_0)]/\lambda$

$$t = (9.16 \times 10^{-13})^{-1} \ln[0.1/(58.6 \times 10^{11})] = -31.7/(-9.16 \times 10^{-13})$$

$t = 3.46 \times 10^{13}$ s = $\underline{\text{1.10 million years}}$. Food for Thought!

---

42. (a) $R = R_0 e^{-\lambda t}$, $R_0 = N_0\lambda = 1.3 \times 10^{-12} N_0(^{12}C)\lambda$

$$R_0 = (1.3 \times 10^{-12} \times 25 \times 6.02 \times 10^{23}/12)\lambda$$

where $\lambda = 0.693/(5730 \times 3.15 \times 10^7) = 3.84 \times 10^{-12}$ decays/s

So R_0 = 376 decays/min , and

$$R = (3.76 \times 10^5)\exp(-3.84 \times 10^{-12} \text{ s}^{-1} \cdot 2.5 \times 10^4 \text{ y} \cdot 3.15 \times 10^7 \text{ s/y})$$

$R = \underline{18.3}$ counts/min

(b) The observed count rate is slightly less than the average background and would be difficult to measure accurately within reasonable counting times.

---

43. The initial activity is $R_0 = 5 \times 10^6$ Ci. The final activity is

$$R_f = (2 \ \mu\text{Ci/m}^2)(10^4 \text{ km}^2) = 2 \times 10^4 \text{ Ci}.$$

$$R_f = R_0 e^{-\lambda t} = R_0 \, 2^{-t/T_{1/2}} = R_0 \, e^{-t \ln 2/T_{1/2}}$$

$$t = (T_{1/2}/\ln 2)\ln(R_0/R_f) = (27.7 \text{ yrs}/0.693)\ln[5 \times 10^6 \text{ Ci}/2 \times 10^4 \text{ Ci})]$$

t = 221 years, not counting leaching.

---

44. Note: Let $M_p = 2.8M_n = 2.8M$

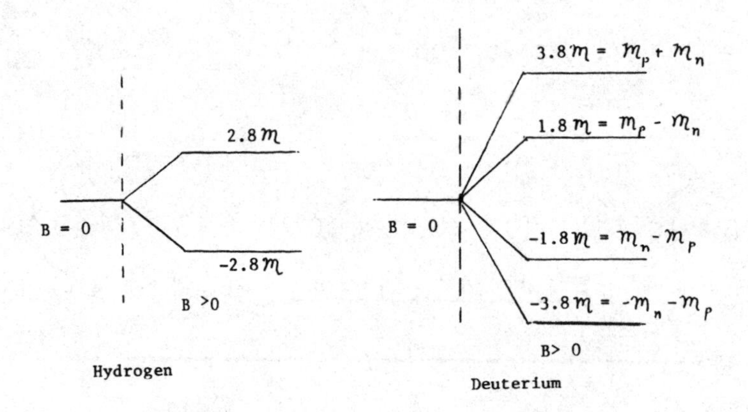

For H: $\Delta E = 2(2.8M)B = 5.6MB_1$

For Deuterium: $\Delta E = 2MB$ (2 transitions)

$\Delta E = 3.6MB$

$\Delta E = 5.6MB$ (2 transitions)

$\Delta E = 7.6MB$

---

45. $R = \lambda N^{147} = \lambda(0.15N^{Sm}) = (\ln 2/T_{1/2})(0.15)(m/M)N_0$

Therefore, the specific activity

$$R/m = (\ln 2/T_{1/2})(0.15)(m/M)(N_0)$$

$$R/m = [0.693/(1.3 \times 10^{10}\text{ years})(3.16 \times 10^7\text{ s/years})]$$

$$\times (0.15)(6.03 \times 10^{23}\text{ atoms/mole}/150.4\text{ g/mole})$$

$R/m = \underline{1.02 \times 10^3}$ Bq/g

---

46. (a) The reaction $p \rightarrow n + \beta^+ + \upsilon$ violates the law of conservation of energy

$$m_p = 1.007276 \text{ u}, \quad m_n = 1.008665 \text{ u}, \quad m_{\beta+} = 5.49 \times 10^{-4} \text{ u}$$

Note that $m_n + m_{\beta+} > m_p$

(b) The required energy can come from the rest energy of the heavier parent nucleus.

(c) $Q = c^2[m(^{13}N) - m(^{13}C) - m_{\beta+} - m_{\upsilon})$

$$Q = (931.5 \text{ MeV/u})(13.005738 - 13.00354 - 5.49 \times 10^{-4} - 0) \text{ u}$$

$$Q = (931.5 \text{ MeV/u})(1.835 \times 10^{-3} \text{ u}) = \underline{1.71} \text{ MeV}$$

---

47. (a) $R = R_0 A^{1/3} = 1.2 \times 10^{-15} A^{1/3}$ m.

When $A = 12$, $R = \underline{2.75 \times 10}^{-15}$ m

(b) $F = k(Z - 1)e^2/R^2 = (9 \times 10^9 \text{ N·m}^2/\text{C}^2)(1.6 \times 10^{-19} \text{ C})^2(Z - 1)/R^2$

When $Z = 6$ and $R = 2.75 \times 10^{-15}$ m, $F = \underline{152}$ N

(c) $U = kq_1q_2/R = k(Z - 1)e^2/R = (9 \times 10^9)(1.6 \times 10^{-19})^2(Z - 1)/R$

When $Z = 6$ and $R = 2.75 \times 10^{-15}$ m, $U = 4.19 \times 10^{-13}$ J $= \underline{2.62}$ MeV

(d) $A = 238$; $Z = 92$, $R = \underline{7.44 \times 10}^{-15}$ m, $F = \underline{379}$ N and

$U = 2.82 \times 10^{-12}$ J $= \underline{17.6}$ MeV

---

48. First find the activity per gram at time $t = 0$

$R_0 = N_0(^{14}C)$, where $N_0(^{14}C) = 1.3 \times 10^{-12} N_0(^{12}C)$; and $N_0(^{12}C) = (m/M)N_a$

Therefore $R_0/m = (\lambda N_a/M)(1.3 \times 10^{-12})$ and the activity after decay at time t will be

$$R/m = (R_0/m)e^{-\lambda t} = (\lambda N_a/M)(1.3 \times 10^{-12})\, e^{-\lambda t} \quad \text{where}$$

$$\lambda = (\ln 2)/T_{1/2} = 2.3 \times 10^{-10} \text{ min}^{-1}$$

When $t = 2000$ years

$$R/m = (2.3 \times 10^{-10}\ \text{min}^{-1}/12\ \text{g/mole})(1.3 \times 10^{-12})(6.03 \times 10^{23}\ \text{mole}^{-1})$$

$$\times\ e^{-(2.3 \times 10^{-10}\ \text{min}^{-1})(2000\ \text{y})(5.26 \times 10^{5}\ \text{min/y})}$$

$R/m = \underline{11.8}\ \text{min}^{-1}\text{g}^{-1}$

---

49. $E = \mathbf{M} \cdot \mathbf{B}$ so the energies are $E_1 = +MB$ and $E_2 = -MB$.

$M = 2.7928M_n$ and $M_n = 5.05 \times 10^{-27}$ J/T

$\Delta E = 2MB = 2 \times 2.7928 \times 5.05 \times 10^{-27} \times 12.5 = 3.53 \times 10^{-25}$ J

$\Delta E = \underline{2.2 \times 10^{-6}}$ eV

---

50. We assume an electron in the nucleus with an uncertainty in its position equal to the nuclear diameter. Choose a typical diameter of 10 fm and from the uncertainty principle we have

$\Delta p \approx h/\Delta x = 6.6 \times 10^{-34}\ \text{J·s}/10^{-14}\ \text{m} = 6.6 \times 10^{-20}$ N·s

Using the relativistic energy-momentum expression

$E^2 = (pc)^2 + (m_0c^2)^2$

we make the approximation that $pc \approx (\Delta p)c >> m_0c^2$ so that

$E \approx pc \approx (\Delta p)c = (6.6 \times 10^{-20}\ \text{N·s})(3 \times 10^8\ \text{m/s}) = \underline{19.8 \times 10^{-12}}$ J

$E \approx \underline{124}$ MeV.

However, the most energetic electrons emitted by radioactive nuclei have been found to have energies of less than 10% of this value.

---

51. (a) At $T = 600 \times 10^6$ K, each carbon nuclei has thermal energy of approximately

$(3/2)kT = (1.5)(8.62 \times 10^{-5}\ \text{eV/K})(600 \times 10^6\ \text{K}) = \underline{7.7 \times 10^4}$ eV

(b) Energy released $= [2m(C^{12}) - m(Ne) - m(He^4)]c^2$

$= (24.000000 - 19.992440 - 4.002603)(931.49)\ \text{MeV} = \underline{4.62}$ MeV

Energy released = $[2m(C^{12}) - m(Mg^{24})](931.49)$ MeV

$= (24.000000 - 23.985042)(931.49)$ MeV = 13.9 MeV

(c) Energy released = the mass of the # of carbon nuclei in a 1 kg sample, which corresponds to

$(1 \text{ kg})(4.62 \text{ MeV})(\text{kW}\cdot\text{h/MeV})$

$= (10^3 \text{ g} \times 6.02 \times 10^{23} \text{ atoms/mole}/12 \text{ g/mole})(4.62 \text{ MeV}) \times [1 \text{ kW}\cdot\text{h}/(2.25 \times 10^{19} \text{ MeV})]$

$= (0.5 \times 10^{26})(4.62)/(2.25 \times 10^{19}) \text{ kW}\cdot\text{h} = \underline{10.3 \times 10^6} \text{ kW}\cdot\text{h}$

---

52. Let N_1 = number of parent nuclei, and N_2 = number of daughter nuclei. The daughter nuclei increase at the rate at which the parent nuclei decrease, or

$$dN_2/dt = -dN_1/dt = \lambda N_1 = \lambda N_{01}e^{-\lambda_1 t}$$

$$dN_2 = \lambda N_{01}e^{-\lambda_1 t}\,dt$$

$$N_2 = \lambda N_{01}\int e^{-\lambda t}\,dt = -N_{01}e^{-\lambda t} + \text{Const.}$$

If we require $N_2 = N_{02}$ when $t = 0$ then Const $= N_{02} + N_{01}$

Therefore $N_2 = N_{02} + N_{01} - N_{01}e^{-\lambda t}$

and when $N_{02} = 0$; $N_2 = N_{01}(1 - e^{-\lambda t})$

---

53. (a) $R = \lambda N = \lambda(m/M)N_A$ or $m = MR/\lambda N_A$

where $N_A = 6.022 \times 10^{23}$ particles/mole is Avogadro's number

and $\lambda = 0.693/T_{1/2} = 0.693/(5.2 \text{ y}) = 0.133 \text{ y}^{-1}$

$\lambda = 4.22 \times 10^{-9} \text{ s}^{-1}$

Also $R = R_0e^{-\lambda t}$ so $R_0 = Re^{\lambda t}$.

Therefore the <u>initial</u> mass of ^{60}Co must be $m = M(Re^{\lambda t})/\lambda N_A$.

Taking $M = 60$ g/mole, $R = (10 \text{ Ci})(3.7 \times 10^{10} \text{ Bq/Ci})$,

$\lambda = (4.22 \times 10^{-9}\ s^{-1})$, and $t = 30$ mo $= 1.89 \times 10^9$ s, we get

$m = \underline{12.2}$ mg

(b) Power = (Energy/decay)(decay/s)

$= [(0.31 + 1.17 + 1.33)\text{MeV/decay}](3.7 \times 10^{11}\ \text{decay/s})$

$= 1.04 \times 10^{12}\ \text{MeV/s} = 0.166\ \text{J/s} = \underline{166}\ \text{mW}$

---

54. The potential at the surface of a sphere of charge q and radiur r is

$$V = kq/r.$$

If a thin shell of charge dq (thickness dr) is <u>added</u> to the sphere, the increase in electrostatic potential energy will be $dU = Vdq = (kq/r)dq$. To build up a sphere with final radius R, the total energy will be

$$U = \int_0^R (kq/r)dq; \quad \text{where}$$

$$q = (4/3)\pi r^3 \rho = (4/3)\pi r^3[Ze/(4\pi R^3/3)] = (Ze/R^3)r^3 \quad \text{so that}$$

$$dq = (3Ze/R^3)r^2 dr$$

$$U = (3kZ^2e^2/R^6)\int_0^R r^5 dr = 3k(Ze)^2/5R$$

(b) When $N = Z = A/2$ and $R = R_0A^{1/3}$, $U = \underline{2.88 \times 10^{-14}}\ (A^{5/3})$ J

(c) For $A = 30$, $U = 8.3 \times 10^{-19}$ J $= \underline{52.1}$ MeV

---

55. We need to find $N/N_0 = e^{-\lambda t}$ where t is the time required to travel a distance $d = 10$ km, and $mv^2/2 = K$; thus we have

$$t = d/v = d/(2K/m)^{1/2} = (m/2K)^{1/2}d$$

Therefore $$N/N_0 = e^{-(\ln 2/T_{1/2})(m/2K)^{1/2}d}$$

$$N/N_0 = e^{-(0.693/720)(10^4)\{1.675 \times 10^{-27}/[2(0.04)(1.6 \times 10^{-19})]\}^{1/2}}$$

$N/N_0 = 0.9965$ or fraction decaying will be $\underline{0.35}$%

---

56. (a) If ΔE is the energy difference between the excited and ground states of the nucleus of mass M, and hf is the energy of the emitted photon, conservation of energy gives

$$\Delta E = hf + E_r \qquad (1)$$

where E_r is the recoil energy of the nucleus, which can be expressed as

$$E_r = Mv^2/2 = (Mv)^2/2M \qquad (2)$$

Since momentum must also be conserved, we have

$$Mv = hf/c \qquad (3)$$

Hence, E_r can be expressed as $E_r = (hf)^2/2Mc^2$

When $hf << Mc^2$, we can make the approximation that $hf \approx \Delta E$, so

$$E_r \approx (\Delta E)^2/2Mc^2$$

(b) $E_r = (\Delta E)^2/2Mc^2$ where $\Delta E = 0.0144$ MeV and

$$Mc^2 = (57\text{ u})(931.5\text{ MeV/u}) = 5.31 \times 10^4\text{ MeV}$$

Therefore

$$E_r = (1.44 \times 10^{-2}\text{ MeV})^2/(2)(5.31 \times 10^4\text{ MeV}) = \underline{1.94 \times 10^{-9}}\text{ MeV}$$

---

57. (a) We will assume the parent nucleus (mass M_p) is initially at rest, and we will denote the masses of the daughter nucleus and alpha particle by M_d and M_α, respectively. The equations of conservation of momentum and energy for the alpha decay process are

$$M_d v_d = M_\alpha v_\alpha \qquad (1)$$

$$M_p c^2 = M_d c^2 + M_\alpha c^2 + (1/2)M_d v_d^2 + (1/2)M_\alpha v_\alpha^2 \qquad (2)$$

The disintegration energy Q is given by

$$Q = (M_p - M_d - M_\alpha)c^2 = (1/2)M_d v_d^2 + (1/2)M_\alpha v_\alpha^2 \qquad (3)$$

Eliminating v_d from Eqs. (1) and (3) gives

$$Q = (1/2)M_d[(M_\alpha/M_d)v_\alpha]^2 + (1/2)M_\alpha v_\alpha^2$$

$$Q = (1/2)(M_\alpha^2/M_d)v_\alpha^2 + (1/2)M_\alpha v_\alpha^2$$

$$Q = (1/2)M_\alpha v_\alpha^2(1 + M_\alpha/M_d)$$

(b) $K_\alpha = Q/(1 + M_\alpha/M) = 4.87\text{ MeV}/(1 + 4/226) = \underline{4.79}\text{ MeV}$

---

58. (a) $r = r_D + r_T = (1.2 \times 10^{-15}\text{ m})(2^{1/3} + 3^{1/3}) = \underline{2.70 \times 10^{-15}}\text{ m}$

(b) $U = ke^2/r = (9 \times 10^9)(1.6 \times 10^{-19})^2/(2 \times 10^{-15})\text{ J}$

$= 1.15 \times 10^{-13}\text{ J} = \underline{720}\text{ keV}$

(c) Conserving momentum: $v_F = v_0 m_D/(m_D + m_T) \qquad (1)$

(d) $(1/2)m_D m_0^2 = (1/2)(m_D + m_T)v_F^2 + U \qquad (2)$

Eliminating v_F from (2) using (1), gives

$$(m_D/2)v_0^2 - (1/2)(m_D + m_T)v_0^2 m_D^2/(m_D + m_T)^2 = U \quad \text{or}$$

$$(1/2)(m_D + m_T)m_D v_0^2 - (1/2)m_D^2 v_0^2 = (m_D + m_T)U \quad \text{or}$$

$$(1/2)m_D^2 v_0^2 = [(m_D + m_T)/m_T]U = (5/3)U = (5/3)720\text{ keV}$$

$$(1/2)m_D^2 v_0^2 = \underline{1.2}\text{ MeV}$$

(c) Possibly by tunnelling.

---

59. (a) $R = R_0 A^{1/3}$, $R_0 = 1.2 \times 10^{-15}\text{ m}$, $R = 10^4\text{ m}$

$A = (R/R_0)^3 = [10^4/(1.2 \times 10^{-15})]^3 = (8.33 \times 10^{18})^3 = \underline{5.8 \times 10^{56}}$

$M = m_n A = (1.67 \times 10^{-27}\text{ kg})(5.8 \times 10^{56}) = \underline{9.7 \times 10^{29}}\text{ kg}$

(about 1/2 the mass of the sun)

(b) $mg = GMm/R^2$ so $g = GM/R^2$

$mg = (6.67 \times 10^{-11}\ N{\cdot}m^2/kg^2)(9.7 \times 10^{29}\ kg)/(10^4\ m)^2 = \underline{6.5 \times 10^{11}}\ m/s^2$

(c) $K = (1/2)I\omega^2$; $\omega = (30 \times 2\pi)$ rad/s; and

$I = (2/5)MR^2 = (2/5)(9.7 \times 10^{29})(10^4)^2 = 3.88 \times 10^{37}\ kg{\cdot}m^2$

so $K = (1/2)(3.88 \times 10^{37})(30 \times 2\pi)^2 = \underline{6.9 \times 10^{41}}\ J$

---

60. A number of atoms, $dN = \lambda N dt$, have life times of t.Therefore, the average or mean life time will be $\Sigma(dN)(t)/\Sigma dN$ or $\int(dN)t/N_0$

$$\text{so} \qquad \tau = (1/N_0)\int_0^\infty \lambda N t\ dt = (1/N_0)\int_0^\infty \lambda N_0 e^{-\lambda t}\ t\ dt = \underline{1/\lambda}$$

---

61. $R_{total} = \lambda N_{(K-40)} = (1.2 \times 10^{-5} N_K)$

$R_{total} = (\ln 2/T_{1/2})(1.2 \times 10^{-5})(m/M)N_a$

$R_{total} = [0.693/(1.27 \times 10^9\ yr)(3.16 \times 10^7\ s/y)](1.2 \times 10^{-5})$

$\times\ (1000\ g/74.55\ g/mole)(6.03 \times 10^{23}\ atoms/mole)$

$R_{total} = 1.677 \times 10^3\ Bq$ and $R_{\beta-} = 0.89 R_{total} = \underline{1.49 \times 10^3}\ Bq$

---

62. At threshold, the particles have no kinetic energy relative to each other. That is, they move like two particles which have suffered a perfectly inelastic collision. Therefore in order to calculate the reaction threshold energy, we can use the results of a perfectly inelastic collision. If M_1 moves with a velocity v_1, and M_2 is initially at rest, we have from momentum conservation

$$M_1 v_1 = (M_1 + M_2)v_c \qquad (1)$$

The initial energy is given by

$$E_i = (1/2)M_1 v_1^2 \qquad (2)$$

The final kinetic energy is

$$E_f = (1/2)(M_1 + M_2)v_c^2 = (1/2)(M_1 + M_2)[M_1 v_1/(M_1 + M_2)]^2 \quad \text{or}$$

$$E_f = (1/2)M_1 v_1^2\, M_1/(M_1 + M_2) \qquad (3)$$

From Eqs. (2) and (3), we see that,

$$E_f = [M_1/(M_1 + M_2)]E_i$$

From this, we see that E_f is always *less* than E_i.

The <u>loss</u> in energy, $E_i - E_f$, is given by

$$E_i - E_f = E_i[1 - M_1/(M_1 + M_2)] = [M_2/(M_1 + M_2)]E_i$$

In this problem, the energy loss equals the disintegration energy $-Q$, the initial energy is the threshold energy E_{th}, $M_1 = M_a$ (the incident particle), and $M_2 = M_X$ (the target), so

$$-Q = [M_X(M_a + M_X)]\, E_{th} \quad \text{or} \quad E_{th} = -Q(1 + M_a/M_X)$$

Note that for an endothermic reaction, Q is negative, so E_{th} is positive.

(b) First calculate the Q-value

$$Q = [m(^{14}N) + m(^{2}He) - m(^{17}O) - m(^{1}H)](931.5 \text{ MeV/u}) = -1.19 \text{ MeV}$$

Then $E_{th} = -Q\,[1 + m(^4He)/m(^{14}N)]$

$$E_{th} = -(-1.19 \text{ MeV})[1 + 4.002603/14.003074] = \underline{1.53} \text{ MeV}$$

---

63. The initial specific activity of ^{59}Fe in the steel, $(R/m)_0$, is

$$(R/m)_0 = (20\ \mu Ci)/(0.2\ kg) = (100\ \mu Ci/kg)(3.7 \times 10^4\ Bq/\mu Ci)$$

$$(R/m)_0 = 3.7 \times 10^6\ Bq/kg$$

After 1000 hr

$$R/m = (R/m)_0 e^{-\lambda t} = (3.7 \times 10^6 \text{ Bq/kg})\, e^{-(6.4 \times 10^{-4} \text{ h}^{-1})(1000 \text{ h})}$$

$$R/m = 1.95 \times 10^6 \text{ Bq/kg}$$

The activity in the oil, $R_{oil} = (800/60)(\text{Bq}/\ell)(6.5\ \ell) = 86.7$ Bq

Therefore

$$m_{in\ oil} = R_{oil}/(R/m) = (86.7 \text{ Bq})/(1.95 \times 10^6 \text{ Bq/kg}) = 4.45 \times 10^{-5} \text{ kg}$$

So that the wear rate = $(4.45 \times 10^{-5}$ kg$)/(1000$ h$) =$ $\underline{4.45 \times 10^{-8}}$ kg/h

---

Chapter 14

1. If N_1 is the particles emerging from the first layer, then

$N_1 = N_0 e^{-\sigma n_1 x_1}$. Also, $N = N_1 = e^{-\sigma n_2 x_2}$ and therefore,

$N = N_0 e^{-\sigma(n_1 x_1 + n_2 x_2)}$. For a material consisting of k layers:

$$N = N_0 \exp -\sigma\left(\sum_{i=1}^{k} n_i x_i\right).$$

---

2. $R = R_0\, e^{-n\sigma x}$, $x = 2$ m, $R = 0.8R_0$

$n = \rho/m_{atom} = (70\ \text{kg/m}^3)/(1.67 \times 10^{-27}\ \text{kg}) = 4.19 \times 10^{28}\ \text{m}^{-3}$

$0.8R_0 = R_0\, e^{-n\sigma x}$, $0.8 = e^{-n\sigma x}$, $n\sigma x = -\ln 0.8$

$\sigma = (-1/nx)\ln(0.8) = 0.223/(4.19 \times 10^{28}\ \text{m}^{-3} \times 2) = 2.66 \times 10^{-30}$ m

$=$ 0.0266 bn

---

3. (a) $I = I_0\, e^{-n\sigma x}$ or

$n\sigma = (1/x)\ln(I_0/I) = (1/1\ \text{cm})\ln(1/0.2865) = 1.25\ \text{cm}^{-1}$

To reduce to 10^{-4}, $x = (1/n\sigma)\ln(10^4) = (\text{cm}/1.25)\ln(10^4) =$ 7.37 cm

(b) $\sigma = (1.25\ \text{cm}^{-1})/n = (1.25\ \text{cm}^{-1})(m/nm) = (1.25\ \text{cm}^{-1})(m/\rho)$

$\sigma = (1.25\ \text{cm}^{-1})(207.2)(1.66 \times 10^{-24}\ \text{g})/(11.35\ \text{g/cm}^3)$

$\sigma = 3.70 \times 10^{-23}\ \text{cm}^2 =$ 38 bn

---

4. Equation 14.1 gives $R = (R_0 n x)\sigma$. Using values of E and σ, we have

$[R_{10}/R_1] = \sigma_{10}/\sigma_1 = 0.0373$, $[R_1/R_{0.1}] = 0.0663$, and $[R_{0.1}/R_{0.01}] \approx 1$

Therefore we can use cadmium as an energy selector in the range 0.1 eV to 10 eV to detect order of magnitude changes in energy.

---

5. σ = 2000 barns. The attenuation length is L, where $n = n_0\, e^{-x/L}$.

From Eq. 14.2, $L = 1/n\sigma = m/\rho\sigma$

$L = (112.41\ \text{u}\cdot\text{cm}^3/[(8.65\ \text{g})(2000\ \text{bn})])(\text{g}/6.02\times10^{23}\ \text{u})(1\ \text{bn}/10^{-24}\ \text{cm}^2)$

$L = \underline{0.01}$ cm

---

6. (a) $N/N_0 = e^{-n\sigma x}$, x = thickness in m, σ = cross section in m^2

and

n = # gold nuclei/m^3

$n = (6.02\times10^{23}\ \text{atoms/mole})(1\ \text{mole}/197\ \text{g})(19.3\ \text{g/cm}^3)$

$n = 5.9\times10^{22}\ \text{atoms/cm}^3 = 5.9\times10^{28}\ \text{atoms/m}^3$

Taking $x = 5.1\times10^{-5}$ m, we get

$N/N_0 = \exp(-5.9\times10^{28}\times500\times10^{-28}\times5.1\times10^{-5}) = \underline{0.86}$

(b) $N = 0.86N_0$ $N_0 = (0.1\ \mu\text{A})/(1.6\times10^{-19}\ \text{C})$

$N_0 = 6.3\times10^{11}$ protons/s and $N = \underline{6.1\times10^{11}}$ protons/s

(c) The number of protons abs. or scatt. per sec = $\underline{8.7\times10^{10}}$ protons/s

---

7. $\Delta E = c^2(m_U - m_{Ba} - m_{Kr} - 2m_n)$

$\Delta E = (931.5\ \text{MeV/u})[235.0439 - 140.9139 - 91.8973 - 2(1.0087)\ \text{u}]$

$\Delta E = (931.5\ \text{MeV/u})[0.2153\ \text{u}] = \underline{200.6}$ MeV

---

8. $(6N_6 + 7N_7)u = 1$ kg where

$N_6/N_7 = 0.074/0.926$ and $u = 1.661\times10^{-27}$ kg

Solve to find $N_7 = 8.05\times10^{25}$

and $N_6 = (0.074/0.926)\times N_7 = 6.43\times10^{24}$

---

9. (a) $(10^9\ W)(1\ day)(24\ h/day)(3600\ s/h)(nucleon/0.9\ MeV)$

$\times (1.66 \times 10^{-24}\ g/nucleon)/(1.6 \times 10^{-13}\ J/MeV) = \underline{996}\ g$

(b) $(4/3)\pi r^3\rho = m;$ diameter $= 2r = (6m/\pi\rho)^{1/3} = \underline{4.87}$ m.

---

10. For a sphere: $V = (4/3)\pi r^3$ and $r = (3V/4\pi)^{1/3}$, so

$A/V = 4\pi r^2/[(4/3)\pi r^3] = 4.84V^{-1/3}$

For a cube: $V = \ell^3$ and $\ell = V^{1/3}$, so $A/V = 6\ell^2/\ell^3 = 6V^{-1/3}$

For a parallelipiped: $V = 2a^3$ and $a = (2V)^{1/3}$, so

$A/V = (2a^2 + 8a^2)/2a^3 = 6.30V^{-1/3}$

Therefore for a given volume, the sphere has the least leakage and the parallelipiped has the greatest leakage.

---

11. t = Energy/Power = 2664 yrs.

---

12. (a) $E = ke^2Z_1Z_2/r = (9 \times 10^9)(1.6 \times 10^{-19})^2Z_1Z_2/10^{-14}$

$= \underline{2.3 \times 10^{-19}}Z_1Z_2$ J

(b) D-D and D-T: $Z_1 = Z_2 = 1$ and $E = 2.3 \times 10^{-19}\ J = \underline{0.14}$ MeV

(c) $Z_1 = 1$ and $Z_2 = 2$, so that $E = 4.6 \times 10^{-19}\ J = \underline{0.28}$ MeV

(d) $Z_1 = Z_2 = 2$, so that $E = 9.2 \times 10^{-19}\ J = \underline{0.56}$ MeV

(e) $Z_1 = 3$ and $Z_2 = 1$ and $E = 6.9 \times 10^{-19}\ J = \underline{0.42}$ MeV.

---

13. (a) Average KE per particle = $(3/2)kT$ $mv^2/2 = (3/2)kT$. Therefore

$v_{rms} = (3kT/m)^{1/2}$

$= [3(1.38 \times 10^{-23}\ J/K)(10^8\ K)/2(1.67 \times 10^{-27}\ kg)]^{1/2}$

$v_{rms} = 1.11 \times 10^6$ m/s;

(b) $t = x/v = (0.1\ m)/(1.11 \times 10^6\ m/s) = \underline{9 \times 10^{-8}}$ s

---

Chapter 14

14. (a) $E = (931.5 \text{ Mev/u})\Delta m = (931.5)[(2\times 2.014102) - 4.002603]$

$E = 23.85$ MeV for every two ^{2}H's.

$(3.17 \times 10^8 \text{ mi}^3)[(5280 \text{ ft/mi})(12 \text{ in/ft})(0.0254 \text{ m/in})]^3$

$\times [10^6 \text{ g}(H^2O)/\text{m}^3][2 \text{ g}(H)/18 \text{ g}(H^2O)][6.02 \times 10^{23} \text{ protons/g}(H)]$

$\times (0.0156\ ^2\text{H/proton})(23.85 \text{ MeV}/^2\text{H})(1.6 \times 10^{-13} \text{ J/MeV})$

$= 2.63 \times 10^{33}$ J,

or $[2.63 \times 10^{33} \text{ J}/(7 \times 10^{14} \text{ J/s})][\text{yr}/(3.16 \times 10^7 \text{ s})] = \underline{119 \text{ billion years}}$

---

15. From Equation 21.6 (Serway PSE) applied to ions and electrons.

(a) $E = 2N(mv^2/2)V = 3NkTV$

$E = 3(10^{14}/\text{cm}^3)(10^4 \text{ eV})(50 \text{ m}^3)(1.6 \times 10^{-19} \text{ J/eV})(10^6 \text{ cm}^3/\text{m}^3)$

$E = \underline{2.4 \times 10^7}$ J

(b) The latent heat of fusion for water is 540 cal/g.

$(2.4 \times 10^7 \text{ J})/[(4.18 \text{ J/cal})(540 \text{ cal/g})(10^4 \text{ g/kg})]$

$= \underline{10.6}$ kg, or $\underline{10.6}$ liters

---

16. (a) $n = (10^{14} \text{ s/cm}^3)/(1 \text{ s}) = \underline{10^{14}/\text{cm}^3}$

(b) $2nkT = (2 \times 10^{14}/\text{cm}^3)(1.38 \times 10^{-23} \text{ J/K})(8 \times 10^7 \text{ K})(10^6 \text{ cm}^3/\text{m}^3)$

$2nkT = \underline{2.2 \times 10^5}$ J/m^3 .

(c) $B^2/2\mu_0 \approx 10(2nkT)$ $\qquad B = [20\mu_0(2nkT)]^{1/2}$

$B = [20(4\pi \times 10^{-7} \text{ N/A}^2)(2.2 \times 10^5 \text{ J/m}^3)]^{1/2} = \underline{2.35}$ T

---

17. $E(x) = hc/\lambda = (6.6 \times 10^{-24})(3. \times 10^{8})/[(25 \times 10^{-12})(1.6 \times 10^{-12})]$

$= 0.050$ MeV

$E(\gamma) = 0.1$ MeV

$E(x)/E(\gamma) = \underline{0.50}$

$\mu(x)/\mu(\gamma) = 55/59.8 = 0.932$

The relationship is not linear since greater change in μ caused by a smaller change in E.

---

18. $x = (\ln 2)/\mu = (\ln 2)/(0.18) = \underline{3.85}$ cm

This means that x-rays can probe the human body to a depth of at least 3.85 cm without severe attenuation and probably farther with reasonable attenuation.

---

19. (a) $I = I_0 e^{-\mu x}$;

$x = (1/\mu)\ln(I_0/I) = (\text{cm}/0.0626)\ln(2000) = 121 \text{ cm} = 1.21 \text{ m}.$

(b) $x = (\text{cm}/0.0626)\ln 10^5 = 183 \text{ cm} = 1.83 \text{ m}.$

---

20. Assume she works 5 days per week, 50 weeks per year and takes 8 x-rays per day .

x-rays = 2000 x-rays per year and 5/2000 = 0.0025 rem per x-ray

5 rem/yr is 38 times the background radiation of 0.13 rem/yr.

---

21. Dose in Rem = (Dose in rad) × RBE

x-rays: (Dose in Rem) = 100 × 1

Heavy Ions: (Dose in Rem) = Z × 20

Therefore 20Z/100 = 1 Z = $\underline{5}$ rads

---

22. The second worker received twice as much radiation energy but he received it in twice as much tissue. Radiation dose is an intensive, not extensive quantity-- measured in joules per kilogram. If you double this energy and the exposed mass, the number of rads is the same in the two cases.

---

23. Rest energy = mc^2 ; N rads applied and

$N(10^{-2}) = mc^2 = (1)(3 \times 10^8) = 9 \times 10^{16}$

$N = \underline{9 \times 10^{18}} \approx 10^{19}$ rads

Thermal energy $\approx \mu T$ for each molecule

molecules = 6.02×10^{23} molecules/mole[1 kg/18 g/mole)](10^3 kg/g)

$= 3.34 \times 10^{25}$

and $N(10^{-2}) = (3.34 \times 10^{25})(1.38 \times 10^{-23}\ J/K)(300\ K)$

where $T = 300$ K at room temperature and

$D = NkT/m = 1.38 \times 10^7$ rads

---

24. One rad → Deposits 10^{-2} J/kg, therefore 25 rad → 25×10^{-2} J/kg

If $M = 75$ kg, $E = (75\ kg)(25 \times 10^{-2}\ J/kg) = \underline{18.8}$ J

---

25. (a) $E/E_\beta = (1/2)CV^2/(0.5\ MeV)$

$= [(1/2)(5 \times 10^{-12}\ F)10^6 V^2/0.5\ MeV]/(1.6 \times 10^{-13}\ J/MeV)$

$= \underline{3.1 \times 10^7}$

(b) $N = Q/e = CV/e = (5 \times 10^{-12}\ F)(10^3\ V)/(1.6 \times 10^{-19}\ C)$

$= \underline{3.1 \times 10^{10}}$ electrons

---

26. One electron strikes the first dynode with 100 eV of energy: 0.10 electrons are freed from the first dynode. These are accelerated to the second dynode . By conservation of energy the number freed here is : $Ze\Delta V = NE_I$

and $10(200 - 100) = N(10)$ so $N = 100$,

By the seventh dynode, $N = 10^6$ electrons . Up to the seventh dynode, we assume all energy is conserved (no losses) . Hence we have 10^6 electrons impinging on the seventh dynode from the sixth . These are accelerated through (700 - 600) V. Hence $E = (10^6)(100) = 10^8$ eV. In addition some energy is needed to cause the 10^6 electrons at the seventh dynode to move to the counter.

---

27. Decay constant $\lambda = 0.693/T = 0.693/(12\ y)(3.16 \times 10^7\ s/y)$

$\lambda = 1.83 \times 10^{-9}\ s^{-1}$

$-\,dN/dt = N\lambda = (1.5 \times 10^{14}/cm^3)(50\ m^3)(1.83 \times 10^{-9}/s)(10^6\ cm^3/m^3)$

$\times$ Ci s/(3.7 $\times$ 10^{10}) = 371 Ci << Fission reactor inventory

---

29. $E_T \equiv E(\text{thermal}) = (3/2)kT = 0.0389$ eV

$E_T = (1/2)^n E$ where $n \equiv$ number of collisions, and

$0.0389 = (1/2)^n(10^6)$.

Therefore $n = 24.6$ or 25 collisions.

---

32. n_e = (# e's/atoms)(atom/mass)(mass/volume)

$n_e = Z \times (6.02 \times 10^{26}/\text{atomic wt.}) \times \text{density}$

	Z	A.W.	$\rho(kg/m^3)$	$n_e(m^{-3})$	μ(0.1 MeV)	μ(10 MeV)
H_2O	10	18 u	1000	3.34×10^{29}	0.167 cm^{-1}	0.0221 cm^{-1}
Al	13	27 u	2700	7.83×10^{29}	0.432	0.0626
Fe	26	55.8 u	7800	21.9×10^{29}	2.69	0.236
Pb	82	207.2 u	11350	27.0×10^{29}	59.8	0.568

Conclusions: The high energy γ's are much more penetrating, and (very roughly) μ is proportional to n_e, the electron density.

---

33. (a) n(slow) + $^6Li \rightarrow {}^3H + {}^4He + Q$

$\Delta m = 1.00866 + 6.01513 - 3.01605 - 4.00260 = 0.00514$ u

$\Delta m = (0.00514\text{ u})(938.3\text{ MeV}/1.00728\text{ u}) = \underline{4.80}$ MeV

(b) n(fast) + $^7Li \rightarrow {}^3H + {}^4He$ + n(slow)

$\Delta m = 7.01600 - 3.01605 - 4.00260 = -0.00265$ u

$\Delta m = -0.00265\ (938.3/1.00728) = \underline{2.45}$ MeV

---

34. For the first layer: $I_1 = I_0 e^{-\mu_{Al} d}$,

for the second layer: $I_2 = I_1 e^{-\mu_{Cu} d}$ and for the third layer:

$I_0/3 = I_2 e^{-\mu_{Pb} d}$ so that $I_0/3 = I_0 e^{-d(\mu_{Al} + \mu_{Cu} + \mu_{Pb})}$ and

$d = \ln 3/(\mu_{Al} + \mu_{Cu} + \mu_{Pb}) = \ln 3/(5.4 + 170 + 610) = \underline{1.4 \times 10^{-3}}$ cm.

If the copper and aluminum are removed, then

$I = I_0\ e^{-(610 \times 1.40 \times 10^{-3})} = \underline{0.426 I_0}$

About 43% of the x-rays get through whereas 33% got through before.

---

35. $E_k = mv^2/2 = p^2/2m = (3/2)kT$ where $k = 1.38 \times 10^{-23}$ J/K,

$T = 300$ K, and $m = 1.67 \times 10^{-27}$ kg

(a) $p = (3mkT)^{1/2} = [(3)(1.67 \times 10^{-27}\text{ kg})(1.38 \times 10^{-23}\text{ J/K})(300\text{ K})]^{1/2}$

$p = \underline{4.55 \times 10^{-24}}$ kg·m/s

(b) $\lambda = h/p = (6.63 \times 10^{-34}\text{ J·s})/(4.55 \times 10^{-24}\text{ kg·m/s}) = 1.46 \times 10^{-10}$ m

$\lambda = 1.46$ **Å** $= \underline{0.146}$ nm.

This is 10^5 times nuclear sizes and about the same size as atomic sizes.

---

36. After the reaction $mv = m'v'$ and $m(mv^2/2) = m(m'v'^2/2)$ and

since 17.6 MeV $= Q_\alpha + Q_n$,

$Q_\alpha = [m_n/(m_\alpha + m_n)]17.6 = [1.009/(4.003 + 1.009)]17.6 = \underline{3.45}$ MeV

$Q_n = [m_\alpha/(m_\alpha + m_n)]17.6 = [4.003/(4.003 + 1.009)]17.6 = \underline{14.1}$ MeV

Since the neutron is uncharged, it cannot be confined by the B field.

Only Q_α can then be used to achieve critical ignition.

---

37. (a) The number of Pu nuclei in 1 kg

$= (6.02 \times 10^{23}\text{ nuclei/mole})(1000\text{ g}/239.05\text{ g/mole})$

The total energy $= 25.2 \times 10^{23}$ nuclei $\times$ 200 MeV $= 5.04 \times 10^{26}$ MeV

$= (5.04 \times 10^{26}\text{ MeV})(4.45 \times 10^{-20}\text{ kWh/MeV})$

$= \underline{22.4 \times 10^6}$ kWh or $\underline{22\text{ million}}$ kWh.

(b) $E = \Delta mc^2$

$= (3.016050 + 2.014102 - 4.002603 - 1.008665\text{ u})(931.8\text{ MeV/u})$

$= 17.6$ MeV for each D-T fusion

(c) $E_n = (\text{Total number of D nuclei})(17.6)(4.45 \times 10^{-20})$

$E_n = (6.02 \times 10^{23})(1000/2.014)(17.6)(4.45 \times 10^{-20}) = \underline{2.34 \times 10^8}$ kWh.

37. (d) E_n = The number of C atoms in 1 kg × 20 eV

$$E_n = (6.02 \times 10^{26}/12.01)(4.1\ \text{eV}/10^6)(4.45 \times 10^{-20}) = \underline{9.14}\ \text{kWh}.$$

---

38. (a) $\sigma = \pi b^2$, the projected area

(b) $b = 2 \times 10^{-3}$ m, $\sigma = \pi(2 \times 10^{-3}\ \text{m})^2 = \underline{1.26 \times 10^{-5}}\ \text{m}^2$

(c) The emergent intensity is $I(x)$, the incident intensity is I_0. Then

$I(x) = I_0 e^{-n\sigma x}$, $V = Ax = (0.5\ \text{m}^2)(2\ \text{m}) = 1\ \text{m}^3$

$n = N/V = (3 \times 10^4)/(1\ \text{m}^3) = 3 \times 10^4\ \text{m}^{-3}$, $x = 2$ m

so $I(x) = I_0 e^{-n\sigma x} = (0.75\ \text{W/m}^2)\, e^{-(3 \times 10^4)(1.26 \times 10^{-5})2}$

$I(x) = (0.75\ \text{W/m}^2)\, e^{-0.756} = (0.75\ \text{W/m}^2)(0.47) = \underline{0.352}\ \text{W/m}^2$

---

39. (a) $Q = [4(1.007825) - 4.002603 - 2(0.00549)](931.5)$ MeV

$= 25.7$ MeV

Therefore $P = 4 \times 10^{26}\ \text{W} = (N/4)(25.7\ \text{MeV/s})(1.6 \times 10^{-13}\ \text{J/MeV})$

and $N = \underline{3.89 \times 10^{38}}$

(b) $\Delta m = Q/c^2 = [25.7\ \text{MeV}/(3 \times 10^8\ \text{m/s})^2](1.6 \times 10^{-13}\ \text{J/MeV})$

$\Delta m = \underline{4.57 \times 10^{-29}}$ kg

4.57×10^{-29} kg are consumed for every 4 protons that fuse.

Therefore, the *total* mass change is

$\Delta M = (4.57 \times 10^{-29})(3.89 \times 10^{38}/4) = \underline{4.44 \times 10^9}$ kg

---

40. (a) A pellet 10^{-4} m in diameter contains $N_0 \times ({}^2{}_1\text{H} + {}^3{}_1\text{H} + 2e)$ where

$N_0 = (6.02 \times 10^{23}/5\ \text{g})(0.2\ \text{g/cm}^3)(4/3)\pi(10^{-4}/2\ \text{m})^3(10^6\ \text{cm}^3/\text{m}^3)$

or $N_0 = 1.26 \times 10^{16}$ particles. With $4N_0$ particles, the temperature is given by $E = (3/2)(4N_0)kT$ or

40. (Cont'd)

$T = E/6N_0k = (0.01)(2 \times 10^{-5}\ J)/(6)(1.26 \times 10^{16})(1.38 \times 10^{-23}\ J/K)$

$T = \underline{1.92 \times 10^9}\ K$

(b) The energy released = $(17.59)(1.26 \times 10^{16})(1.6 \times 10^{-13}) = \underline{355}\ kJ$

---

41. (a) $E = mv^2/2$ and $(500 \times 1.6 \times 10^{-19}) = (2 \times 1.67 \times 10^{-27})v^2/2$

so $v = 2.19 \times 10^5$ m/s; $v_{para} = v\cos 30° = \underline{1.90 \times 10^5}$ m/s; and

$v_{perp} = v\sin 30° = \underline{1.10 \times 10^5}$ m/s

(b) From Equation 14.15: $r = mv_{perp}/qB$ and

$r = (2 \times 1.67 \times 10^{-27})(1.09 \times 10^5)/(1.67 \times 10^{-27})(1)$

$= \underline{2.28 \times 10^{-3}}$ m

(c) From Equation 14.17: $T = 2\pi m/qB$ and

$T = 2\pi(2 \times 1.67 \times 10^{-27})/(1.6 \times 10^{-19})(1) = 1.31 \times 10^{-7}$ s and

distance traveled = $v_{para}T = \underline{2.49 \times 10^{-2}}$ m

---

42. The neutron absorption rate to the cadmium is $(1/n)dn/dt = -n_c\sigma v_{th}$ where n = neutron density, n_c = density of cadmium nuclei, σ = is the absorption cross-section, and v_{th} is the neutron thermal velocity given by $v_{th} = [(3/2)(kT/m_n)]^{1/2}$. The neutron decay rate is $(1/n)dn/dt = -0.693/T_{1/2}$, where $T_{1/2} = 10.60$ min = 636 s. The ratio is

$R = n_c\sigma v_{th}\ T_{1/2}/0.693 = (8.65\ g/cm^3)(6.02 \times 10^{23}\ nuclei/112.4\ g)$

$\times\ (2450\ barns/nucleus)(10^{-24}\ cm^2/barn)$

$\times\ [3(1.38 \times 10^{-23}\ J/K)(300\ K)/(1.66 \times 10\ kg)]^{1/2}$

$\times\ (100\ cm/m)(720\ s/0.693) = \underline{2.8 \times 10^{10}}$

---

43. $m(^1H) = 1.007825$ u, $m(^2H) = 2.014102$ u, $m(^3He) = 3.016030$ u,

$m(^4He) = 4.002603$ u

$m_{e^\pm} = 0.000549$ u $\quad m_\upsilon = 0 \quad m_\gamma = 0$

(1) $^1H + {}^1H \rightarrow {}^2H + e^+ + \upsilon$

$$Q_1 = [2m(^1H) - m(^2H) - m_{e^+}]c^2$$

$$Q_1 = [2(1.007825) - 2.014102 - 0.000549 \text{ u}](931.5 \text{ MeV/u})$$

$$Q_1 = (0.00099 \text{ u})(931.5 \text{ MeV/u}) = \underline{0.931} \text{ MeV}$$

(2) $^1H + {}^2H \rightarrow {}^3He + \gamma$

$$Q_2 = [m(^1H) + m(^2H) - m(^3He)]c^2$$

$$Q_2 = (1.007825 + 2.014102 - 3.016030 \text{ u})(931.5 \text{ MeV/u})$$

$$Q_2 = (0.005897 \text{ u})(931.5 \text{ MeV/u}) = \underline{5.493} \text{ MeV}$$

(3) $^1H + {}^3He \rightarrow {}^4He + e^+ + \upsilon$

$$Q_3 = [m(^1H) + m(^3He) - m(^4He) - m_{e^+})c^2$$

$$Q_3 = (1.007825 + 3.016030 - 4.002603 - 0.000549)(931.5) \text{ MeV}$$

$$Q_3 = (0.020703)(931.5) \text{ MeV} = \underline{19.285 \text{ MeV}}$$

(4) $^3He + {}^3He \rightarrow {}^4He + {}^1H + {}^1H$

$$Q_4 = [2m(^3He) - m(^4He) - 2m(^1H)]c^2$$

$$Q_4 = [2(3.016030) - 4.002603 - 2(1.007825) \text{ u}](931.5 \text{ MeV/u})$$

$$Q_4 = (0.013807 \text{ u})(931.5 \text{ MeV/u}) = \underline{12.861 \text{ MeV}}$$

The total reaction is $4{}^1H \rightarrow {}^4He + 2e^+ + 2\upsilon$ which is equivalent to (1) + (2) + (3) or to (2) × (3) − (4)

$$Q_T = Q_1 + Q_2 + Q_3 = 2Q_3 - Q_4 = [4m(^1H) - m(^4He) - 2m(e+)]c^2$$

$$Q_T = 0.931 \text{ MeV} + 5.493 \text{ MeV} + 19.285 \text{ MeV} = 25.709 \text{ MeV}$$

43. (Cont'd)

$Q_T = 2(19.285 \text{ MeV}) - 12.861 \text{ MeV} = 25.709 \text{ MeV } (2Q_3 - Q_4)$

$Q_T = (931.5 \text{ MeV/u})(4)(1.007825) - 4.002603 - (2)(0.000549) \text{ u}$

$Q_T = (931.5 \text{ MeV/u})(0.027599 \text{ u}) = \underline{25.709} \text{ MeV}$

---

44. (a) Roughly $(7/2)(15 \times 10^6)$ K or 52×10^6 K since 6 × the coulombic barrier must be surmounted.

(b) $E = \Delta mc^2 = (12.00000 + 1.007825 - 13.005738 \text{ u})(931.5 \text{ MeV/u})$

$E = 1.943$ MeV

$1.943 + 1.709 + 7.551 + 7.297 + 2.242 + 4.966 = 25.75$ MeV

The net effect is ${}^{12}_{6}C + 4p \rightarrow {}^{12}_{6}C + {}^{4}_{2}He$

(c) Most of the energy is lost since υ's have such low cross-section (no charge, little mass, etc.)

---

45. While the silver is in the reactor, the decay rate of an activated isotope equals its production rate, i.e., $n_n n_{107} \sigma_{107} v_{th} = \lambda_{10} n_{108}$ and $n_n n_{109} \sigma_{109} v_{th} = \lambda_{110} n_{110}$ where λ is the decay rate. At the instant of removal the ratio of ^{108}Ag to ^{110}Ag can be found from the ratio of the two equations above:

$$n_{108}/n_{110} = (n_{107}/n_{109})(\sigma_{107}/\sigma_{109})(\lambda_{110}/\lambda_{108})$$

After a delay t, the densities are

$n'_{108} = n_{108}\, e^{-\lambda_{108} t}$ and $n'_{110} = n_{108}\, e^{-\lambda_{110} t}$, so

$n'_{108}/n'_{110} = (n_{108}/n_{110})e^{-(\lambda_{108} - \lambda_{110})t}$

$n'_{108}/n'_{110} = (n_{107}/n_{109})(\sigma_{107}/\sigma_{109})(\lambda_{110}/\lambda_{108})e^{-(\lambda_{108} - \lambda_{110})t}$ or

$t = [1/(\lambda_{108} - \lambda_{110})]\ln[(n_{107}/n_{109})(\sigma_{107}/\sigma_{109})(\lambda_{110}/\lambda_{108})(n'_{110}/n'_{108})]$

45. (Cont'd)

$t = (-sec/0.0235)\ln[(0.5135/0.4865)(31/87)(144/24.5)(1/20)]$

$t = \underline{95.2}$ s

where $\lambda_{108} = 0.693/(144\ s) = 0.00481\ s^{-1}$ and

$\lambda_{110} = 0.693/(24.55\ s) = 0.02829\ s^{-1}$

---

46. (a) $eff = P_{delivered}/P_{out} = 0.3$

$P_{out} = 1000/0.3 = \underline{3333}$ MW

(b) $P_{heat} = P_{out} - P_{delivered} = 3333 - 1000 = \underline{2333}$ MW

(c) The energy released per fission event is $Q = 200$ MeV. Therefore

Rate $= P_{out}/Q = (3.333 \times 10^9\ W/200\ MeV)/(1.6 \times 10^{-13}\ J/MeV)$

Rate $= \underline{1.04 \times 10^{20}}$ events/s.

(d) $M = R[235 \times 10^{-3}$ kg/mole/$(6.0 \times 10^{23}$ atoms/mole)](time)

$M = (1.04 \times 10^{20})(3.92 \times 10^{-25}$ kg/atom)(365 days)(24 h/day)

$\times$ (3600 s/h) $= \underline{1.34 \times 10^3}$ kg

(e) $dM/dt = (1/c^2)(dE/dt)$

$= (3.333 \times 10^9\ W)/(3 \times 10^8\ m/s)^2 = \underline{3.7 \times 10^{-8}}$ kg/s

---

Chapter 15

1. The minimum energy is released, and hence the minimum frequency photons are produced, when the proton and antiproton are at rest when they annihilate. That is, $E = E_0$, and $K = 0$. To conserve momentum, each photon must carry away one-half the energy. Thus,

$$E_{min} = hf_{min} = (2E_0)/2 = E_0 = 938.3 \text{ MeV}$$

Thus,

$$f_{min} = \frac{(938.3 \text{ MeV})(1.6 \times 10^{-13} \text{ J/MeV})}{6.63 \times 10^{-34} \text{ J·s}} = 2.26 \times 10^{23} \text{ Hz}$$

and $$\lambda = \frac{c}{f_{min}} = \frac{3 \times 10^8 \text{ m/s}}{2.26 \times 10^{23} \text{ Hz}} = \underline{1.32 \times 10^{-15}} \text{ m}$$

---

2. Assuming that the proton and antiproton are left at rest after they are produced, the energy of the photon, E, must be

$$E = 2E_0 = 2(938.3 \text{ MeV}) = 1876.6 \text{ MeV} = 3.00 \times 10^{-10} \text{ J}$$

Thus, $$E = hf = 3.00 \times 10^{-10} \text{ J}$$

and $$f = \frac{3.00 \times 10^{-10} \text{ J}}{6.63 \times 10^{-34} \text{ J·s}} = \underline{4.53 \times 10^{23}} \text{ Hz}$$

and $$\lambda = \frac{c}{f} = \frac{3 \times 10^8 \text{ m/s}}{4.53 \times 10^{23} \text{ Hz}} = \underline{6.62 \times 10^{-16}} \text{ m}$$

---

3. The rest energy of the Z^0 boson is $E_0 = 96$ GeV.

The maximum time a virtual Z^0 boson can exist is found from

$\Delta E \Delta t = \hbar$, or

$$\Delta t = \frac{\hbar}{\Delta E} = \frac{1.055 \times 10^{-34} \text{ J·s}}{(96 \text{ GeV})(1.6 \times 10^{-10} \text{ J/GeV})} = 6.87 \times 10^{-27} \text{ s}$$

The maximum distance it can travel in this time is

$$d = c(\Delta t) = (3 \times 10^8 \text{ m/s})(6.87 \times 10^{-27} \text{ s}) = \underline{2.06 \times 10^{-18}} \text{ m}$$

The distance d is an approximate value for the range of the weak interaction.

---

4. $$\mu^+ + e \rightarrow \upsilon + \upsilon$$

muon-lepton number before reaction = (-1) + (0)

electron-lepton number before reaction = (0) + (1) = 1

Therefore, after the reaction, the muon-lepton number must be -1. Thus, one of the neutrinos must be the anti-neutrino associated with muons or $\bar{\upsilon}_\mu$. Also, after the reaction, the electron-lepton number must be 1. Thus, one of the neutrinos must be the neutrino associated with electrons, or υ_e.

Thus, $$\mu^+ + e \rightarrow \bar{\upsilon}_\mu + \upsilon_e$$

---

5. The time for a particle traveling with the speed of light to travel a distance of 3×10^{-15} m is

$$\Delta t = \frac{d}{v} = \frac{3 \times 10^{-15}\ \text{m}}{3 \times 10^{8}\ \text{m/s}} = 10^{-23}\ \text{s}$$

---

6. The $\rho_0 \to \pi^+ + \pi^-$ decay must occur via the strong interaction.

The $K^0 \to \pi^+ + \pi^-$ decay must occur via the weak interaction.

---

7. (a) $\Lambda^0 \to p + \pi^-$

Strangeness: $-1 \to 0 + 0$ (strangeness is not conserved)

(b) $\pi^- + p \to \Lambda^0 + K^0$

Strangeness: $0 + 0 \to -1 + 1$ (0 = 0 and strangeness is conserved)

(c) $\bar{p} + p \to \bar{\Lambda}^0 + \Lambda^0$

Strangeness: $0 + 0 \to +1 - 1$ (0 = 0 and strangeness is conserved)

(d) $\pi^- + p \to \pi^- + \Sigma^+$

Strangeness: $0 + 0 \to 0 - 1$ (0 does not equal -1 so strangeness is not conserved)

(e) $\Xi^- \to \Lambda^0 + \pi^-$

Strangeness: $-2 \to -1 + 0$ (-2 does not equal -1 so strangeness is not conserved)

(f) $\Xi^0 \to p + \pi^-$

Strangeness: $-2 \to 0 + 0$ (-2 does not equal 0 so strangeness is not conserved)

8. (a) $\mu^- \rightarrow e + \gamma$

L_e: $0 \rightarrow 1 + 0$ and L_μ: $1 \rightarrow 0 + 0$

(b) $n \rightarrow p + e + \nu_e$

L_e: $0 \rightarrow 0 + 1 + 1$

(c) $\Lambda^0 \rightarrow p + \pi^0$

Strangeness: $-1 \rightarrow 0 + 0$, and charge: $0 \rightarrow +1 + 0$

(d) $p \rightarrow e^+ + \pi^0$

baryon number: $+1 \rightarrow 0 + 0$

(e) $\Xi^0 \rightarrow n + \pi^0$

Strangeness: $-2 \rightarrow 0 + 0$

---

9. (a) $p + \overline{p} \rightarrow \mu^+ + e$

L_e: $0 + 0 \rightarrow 0 + 1$ and L_μ: $0 + 0 \rightarrow -1 + 0$

(b) $\pi^- + p \rightarrow p + \pi^+$

charge: $-1 + 1 \rightarrow +1 + 1$

(c) $p + p \rightarrow p + \pi^+$

baryon number: $1 + 1 \rightarrow 1 + 0$

(d) $p + p \rightarrow p + p + n$

baryon number : $1 + 1 \rightarrow 1 + 1 + 1$

(e) $\gamma + p \rightarrow n + \pi^0$ charge: $0 + 1 \rightarrow 0 + 0$

---

11. (a) $\pi^- \to \mu^- + \bar{\upsilon}_\mu$ $\quad$ L_μ: $0 \to 1 - 1$

(b) $K^+ \to \mu^+ + \upsilon_\mu$ $\quad$ L_μ: $0 \to -1 + 1$

(c) $\bar{\upsilon}_e + p \to n + e^+$ $\quad$ L_e: $-1 + 0 \to 0 - 1$

(d) $\upsilon_e + p \to n + e$ $\quad$ L_e: $1 + 0 \to 0 + 1$

(e) $\upsilon_\mu + n \to p + \mu^-$ $\quad$ L_μ: $1 + 0 \to 0 + 1$

(f) $\mu^- \to e + \bar{\upsilon}_e + \upsilon_\mu$ $\quad$ L_μ: $1 \to 0 + 0 + 1$, and

L_e: $0 \to 1 - 1 + 0$

---

12. (a) $p \to \pi^+ + \pi^0$ $\quad$ baryon number is violated: $1 \to 0 + 0$

(b) $p + p \to p + p + \pi^0$ $\quad$ This reaction can occur.

(c) $p + p \to p + \pi^+$ $\quad$ baryon number is volated: $1 + 1 \to 1 + 0$

(d) $\pi^+ \to \mu^+ + \upsilon_\mu$ $\quad$ This reaction can occur.

(e) $n \to p + e + \bar{\upsilon}_e$ $\quad$ This reaction can occur.

(f) $\pi^+ \to \mu^+ + n$ $\quad$ Violates baryon number: $0 \to 0 + 1$, and violates muon-lepton number: $0 \to -1 + 0$

---

13. (a)

	proton	u	u	d	total
strangeness	0	0	0	0	0
baryon number	1	1/3	1/3	1/3	1
charge	e	2e/3	2e/3	-e/3	e

(b)

	neutron	u	d	d	total
strangeness	0	0	0	0	0
baryon number	1	1/3	1/3	1/3	1
charge	0	2e/3	-e/3	-e/3	0

14. (a)

	$\overline{K^0}$	d	s	total
strangeness	1	0	1	1
baryon number	0	1/3	-1/3	0
charge	0	-e/3	e/3	0

(b)

	Λ^0	u	d	s	total
strangeness	-1	0	0	-1	-1
baryon number	1	1/3	1/3	1/3	1
charge	0	2e/3	-e/3	-e/3	0

Part II

Software Instructions

Software Instructions

INTRODUCTION

The program disk which is avaiable upon request from the publisher contains five programs for IBM PC compatible machines running MS DOS. The complete disk contents are listed below:

WAVES-1.EXE	WAVES-1.INI
WAVES-2.EXE	WAVES-2.INI
EIGEN.EXE	EIGEN.INI
EIGEN3.EXE	EIGEN3.INI
ORBS.EXE	ORBS.INI
MP.LIB	
README.DOC	

The file **MP.LIB** is a library of routines supporting the executible programs **xxxx.EXE.** The companion file **xxxx.INI** to each **xxxx.EXE** program specifies defaults for the parent program; its contents may be altered with the **CUSTOMIZE** command described below in **GETTING HELP**. **README.DOC** contains program updates too recent to be included in this manual.

Each executible program is fully interactive, with numerous options allowing the student wide flexibility in exploring the phenomena under study. Menus guiding the user through every step of program execution facilitate the selection process. Help menus inform the student of options when keyboard input is required.

The programs run on any IBM PC compatible machine having at least 512K of memory and an IBM Color Graphics Adapter (CGA). However, an Enhanced Graphics Adapter (EGA) with color monitor is preferred.

GETTING HELP

For assistance with any **xxxx.EXE** program, press the help key [F1]. This displays the **HELP BANNER** across the top of the screen showing the available **HELP FUNCTIONS**, and opens the **COMMAND MENU** for the function highlighted. Move right (or left) among functions by pressing the Right (Left) direction key. To **CLOSE HELP** press the [Esc] key.

The **HELP FUNCTIONS** are:

CONTROL- directs program control as indicated by the CONTROL MENU.

CHANGE- changes program parameters listed in the CHANGE MENU.

ANALYZE- permits quantitative analysis of on-screen graphics. The analysis function is supported by the commands listed in the ANALYZE MENU.

SPECIAL- directs special program operations, as listed in the FEATURES MENU.

The **COMMAND MENU** for each function appears below.* To invoke any command, press [Enter] with the command highlighted, or use the shortcut key listed for that entry. Proceed through the **COMMAND MENU** by pressing the [Dn] or [Up] direction key.

CONTROL MENU

Command	Action	Shortcut Key
Restart	execute program from beginning	Ctrl+ [Home]
Quit	terminate program execution	Ctrl+ [End]
Previous	restore previous screen page	[PgUp]
Customize. . .	make changes to **xxxx.INI** file	

* Menus may vary among programs; those shown are for **WAVES-1.EXE**.

CHANGE MENU

Command	Action	Shortcut Key
Time. . .	select new t	
Mass	input new m	Alt+ [m]
Spectrum. . .	make changes to a(k)	
Update	update graphics display to reflect changes	Ctrl+ [Enter]

ANALYZE MENU

Command	Action	Shortcut Key
Fine	move reticle left|right one unit	[Left] or [F7]+ . . .[Right] or [F8]
Coarse	move reticle left|right ten units	Shift+ [Left] . . .Shift+ [Right]
Limits	move reticle to left|right view limit	Ctrl+ [Left] . . .Ctrl+ [Right]

(The above keys activate the analyzer, producing on-screen readout of the graphics display)

Cancel	delete on-screen readout and exit analyzer	[Home] or [KP5]

([Home] zeroes the reticle; with [KP5], the last reticle position is "remembered" when the analyzer is next activated)

FEATURES MENU

<table>
<tr><th>Command</th><th>Action</th><th>Shortcut Key</th></tr>
<tr><td>Zoom</td><td>magnify | compress graphics field</td><td>[>]
. . .[<]</td></tr>
<tr><td>Pan</td><td>move up | down in graphics field</td><td>Ctrl+ [Up]
. . .Ctrl+ [Dn]</td></tr>
<tr><td>Review</td><td>invoke/toggle inset showing units, etc.</td><td>[F2]</td></tr>
<tr><td>Split</td><td>enter split-screen mode</td><td>[Ins]</td></tr>
</table>

Executing a menu command with trailing ellipses (. . .) opens the **OPTION MENU** listing options for that command. For example, invoking the TIME command from the **CHANGE MENU** displays the **TIME CHANGE MENU.**

Time ahead	[Up] or [F5]*
Time back	[Dn] or [F6]
Cursor left	[Bksp]
Cursor right	[Spacebar]
Start line	[Home]
End line	[End]
Unit	[Tab]

To close any **OPTION MENU** and return to the **COMMAND MENU**, press the [Esc] key.

* The usual shortcut assignments for the direction keys are suspended during the **HELP** call to free these keys for the **HELP** activity. However, commands associated with these keys still may be executed directly from **HELP** by pressing Alt+ [direction key] on the dedicated cursor keypad, or the function keys [F5]-[F8] as noted.

An overview of each executible program follows:

WAVES-1.EXE

HIGHLIGHTS

WAVES-1.EXE calculates and displays wave groups $G(x,t)$ representing [the *real part* of] the matter wave $\Psi(x,t)$ formed by superposing free particle plane waves:

$$G(x,t) = \mathrm{Re}\{\Psi(x,t)\} = \mathrm{Re}\left\{\sum a(k)e^{i(kx-\omega t)}\right\}$$

The *spectral function* $a(k)$ is chosen from listed options, and displayed in the form of a line graph for wavenumbers k in the range specified.

The group $G(x,t)$ is displayed as a function of x at an instant of time t. Provision exists to locate any point on the wave (crest, node, etc.) and follow it as time progresses. In this way accurate measurements can be made of the phase velocity. Similarly, the concepts of group velocity and dispersion are amenable to quantitative investigation. Finally, comparative studies of the spectral function $a(k)$ with $G(x,t)$ allow confirmation of the uncertainty principle governing these waves.

SPECIFICS

The UNIT MENU sets the display range, and fixes the system of units used throughout. Velocities always appear in units of c, the speed of light. The *dispersion relation* for [relativistic] de Broglie waves appears in these units as

$$\omega(k) = (k^2 + m^2)^{1/2}$$

where m is the particle mass.

The spectral function a(k) is chosen from the **SPECTRUM MENU**:

Uniform-	$a(k) = 1$
Harmonic-	$a(k) = \cos(\gamma k\pi)$
Gaussian-	$a(k) = e^{-\alpha^2(k - g)^2}$

Values for the modulus $|a(k)|$ are automatically scaled to a maximum equal to unity.

Wavenumbers used to contruct the group are chosen by specifying their range and increment. The number of group components is limited only by practical considerations (computing time, numerical accuracy, etc). If the maximum and minimum wavenumbers are equal, the program displays the single oscillation with that wavenumber. Setting the increment δk to zero results in *integration* over the range specified.

The **SPECTRAL DISPLAY** is a line graph of a(k) versus k. Negative values for a(k) appear on this graph as broken lines. The average wavenumber (weighted according to $|a(k)|^2$) is marked by a short vertical line (red on color monitor) on the graph abcissa. Values for k, a(k) and $\omega(k)$ for each spectral component may be read from the screen using the **ANALYZE** function. The spectral display is bypassed for a group consisting of a single oscillation.

The **GROUP DISPLAY** is a graph of G(x,t) versus x for some t. For t = 0 (the default), the group is built up one component at a time, each appearing in the **SPECTRAL INSERT** as it is added to the group. The group also may be examined with the **ANALYZE** function, allowing the value for G(x,t) to be read from the screen at any position x. Other values for t are set using the options from the **TIME CHANGE MENU**; pressing Ctrl+ [Enter] refreshes the graphics display. The first call to nonzero t prompts the user to input a mass value m. Further changes to m, or alterations to the spectral content of the group, are effected with the commands listed in the **CHANGE MENU**.

************* SPECIAL FEATURE **************

The spectral function a(k) can be displayed simultaneously with the wave group G(x,t) in a split-screen mode. The split-screen mode is invoked/toggled by executing the **SPLIT** command from the **FEATURES MENU**, or by pressing the [Ins] key. In split-screen mode, the **ANALYZE** commands affect the active viewport only. Pressing [F10] interchanges the active and passive viewports. The split-screen mode is especially useful in comparing the width of a wave group with that of its spectral content to establish uncertainty relations.

****

WAVES-2.EXE

HIGHLIGHTS

WAVES-2.EXE calculates and displays [the real part of] the matter wave group formed as a superposition of stationary states for a chosen quantum potential:

$$G(x,t) = \mathrm{Re}\left\{\sum a(n)\psi_n(x)e^{-iE_n t/\hbar}\right\}$$

The *basis states* $\psi_n(x)$ and their energies E_n are available for several pre-selected potential wells $U(x)$ of widespread interest. For added flexibility, all physical quantities are expressed in generalized units.

The potential energy $U(x)$ is displayed over an interval [a, b] set by the user; values for $U(x)$ may be read from the screen at any point x in this interval.

The *group envelope* $a(n)$ is chosen from listed options, and displayed as a line graph for the state labels n in the range specified.

The wave group $G(x,t)$ is displayed at an instant of time t; its evolution may be tracked by changing t. On-screen readout of the group value $G(x,t)$ at any point x and any time t allows quantitative investigation of its evolution.

SPECIFICS

The basis functions $\psi_n(x)e^{-iE_nt/\hbar}$ comprising the group are stationary states for the force acting on the particle. This force is identified by its associated potential energy entry in the **ENVIRON MENU**:

Infinite Well——— $U(x) = 0$ on [0,1]
$= \infty$ elsewhere

Divided Well——— $U(x) = S\delta(x - 1/2)$ on [0,1]
$= \infty$ elsewhere

Oscillator——— $U(x) = x^2$

The expressions for U(x) are given in generalized units. For example, a natural length for the infinite well is its width L, which then becomes the length unit for this program option. Other generalized program units are similarly identified. Program units may be reviewed by pressing [F2].

The envelope function a(n) is selected from the **SPECTRUM MENU**:

Uniform ——— $a(n) = 1$

Harmonic——— $a(n) = \cos(\gamma n\pi)$

Geometric——— $a(n) = \gamma^n$

The composition of the group is completed by specifying the range and increment of the stationary state labels n. Program convention assigns n = 0 to the ground state, followed by n = 1, 2, 3, . . . for the first, second, third, . . . excited states, respectively. The number of group components is limited only by practical considerations (computing time, numerical accuracy, etc). To display the single stationary wave $\psi_n(x)$, set the maximum and minimum state labels equal to n.

The **SPECTRAL DISPLAY** is a line graph of a(n) versus n. Negative values for a(n) appear on this graph as broken lines. The average energy <E> for the group is marked by a short vertical line (red on color monitor) on the graph abcissa. Values for n, a(n) and E_n (natural units) for each component wave may be read from the screen with the **ANALYZE** function. The spectral display is bypassed for a group composed of a single stationary wave.

The **POTENTIAL DISPLAY** plots U(x) over the **VIEW RANGE** [a,b]. Pressing Alt+ [V] overrides program defaults for the view limits a,b and invites user input. After input is completed, pressing Ctrl+ [Enter] refreshes the graphics display. On-screen readout of U(x) at any point x within the **VIEW RANGE** is accomplished with the **ANALYZE** function.

The **GROUP DISPLAY** is a graph of G(x,t) versus x for some t. The group G(x,t) is generated and studied as described in the synopsis for **WAVES-1**.

Note: The wave group G(x,t) is constructed from *unnormalized* basis functions $\psi_n(x)$ and so is itself unnormalized, even at t = 0. Further, in calculating G(x,t), the energies E_n are referenced to the energy of the lowest lying state in the group, thereby eliminating at the outset any [inconsequential] multiplicative factor $e^{-iEt/\hbar}$.

************* SPECIAL FEATURE **************

The envelope function a(n) can be displayed simultaneously with the wave group G(x,t) in a split-screen mode. See the WAVES-1 synopsis for details regarding the use of this feature.

****

EIGEN.EXE

HIGHLIGHTS

EIGEN.EXE solves the *eigenvalue problem* posed by the [time-independent] Schrodinger equation in one dimension. The allowed energies and stationary states that result are displayed for quantitative on-screen study.

The potential energy U(x) is selected from a menu of seven wells and barriers, many of which include adjustable parameters. For added flexibility, menu entries are given in generalized units. Finally, a user-defined option for U(x) renders the scope of the program virtually limitless.

The potential energy U(x) is displayed on an interval [a, b] set by the user; values for U(x) may be read from the screen at any point x in this interval.

The *eigenvalues* are found by an easy-to-use, interactive trial and error procedure, culminating in a visual display of the associated eigenfunctions on [a, b]. On-screen readout allows full quantitative investigation of the waveform. In the case of *scattering states* , the incident, reflected, and transmitted waves all may be displayed simultaneously, and their amplitudes compared to find coefficients of reflection and transmission.

SPECIFICS

The equation actually solved by **EIGEN**

$$\left\{- d^2/dx^2 + U(x)\right\} f(x) = E\, f(x)$$

is Schrodinger's time-independent equation in natural units for distance and energy. These units ε and δ are interrelated as $\varepsilon\delta^2 = \hbar^2/2m$. For example, adopting the lowest energy of the infinite well, $\pi^2\hbar^2/2mL^2$, as the energy unit ε for all square wells gives a natural length unit $\delta = L/\pi$.

The **ENVIRON MENU** presents the following choices for U(x):

Square Well-	$U(x) = 0$	on $[-\pi/2, \pi/2]$
	$= U$	elsewhere
Triangular Well——	$U(x) = \lvert x \rvert$	
Oscillator——————	$U(x) = x^2$	
Double Oscillator-	$U(x) = (\lvert x \rvert - b)^2$	
Square Barrier———	$U(x) = 1$	on [0, L]
	$= 0$	elsewhere
Tilted Step—————	$U(x) = 0$	on $[-\infty, 0]$
	$= 1 - F \cdot x$	on [0,]
Tilted Barrier————	$U(x) = 1 - F \cdot x$	on [0, L]
	$= 0$	elsewhere
External——————	U(x) is user-defined	

Entries for U(x) are given in natural units; these units may be reviewed during program execution by pressing [F2].

The **POTENTIAL DISPLAY** shows U(x) over the **VIEW RANGE** [a, b]. On-screen readout of U(x) anywhere within the **VIEW RANGE** is accomplished with the **ANALYZE** function. Pressing Alt+ [V] overrides program defaults for the view limits and invites user input. Pressing Alt+ [U] initiates screen prompts for new values of any parameters appearing in U(x). All changes are registered on the graphics display by pressing Ctrl+ [Enter].

A first guess for E is set at the **POTENTIAL DISPLAY** using the [Up]/[Dn] keys. Pressing Ctrl+ [Enter] initiates computation and generates graphics output for f(x) at the **WAVEFUNCTION DISPLAY**. Any discontinuity in f(x) indicates this trial waveform is not an acceptable eigenfunction, and necessitates corrections to E.

Adjustments to E are made using commands from the **ENERGY CHANGE MENU**, and the procedure repeated as necessary until no discontinuity in f(x) is evident. Invoking the **ANALYZE** function produces on-screen readout of the resulting eigenfunction for further study. The **ZOOM** and **PAN** commands from the **FEATURES MENU** facilitate inspection of waveforms exceeding the vertical limits of the graphics viewport. (At the **POTENTIAL DISPLAY** the target of **ZOOM** and **PAN** are energies instead of waveforms, but the effects are similar.)

************* SPECIAL FEATURES **************

For scattering states, only the *real part* of the waveform is shown at the **WAVEFUNCTION DISPLAY**. Pressing [Ins] invokes the **DECON** command from the **FEATURES MENU** and deconvolutes the leftmost portion into incident and reflected components (real parts, again!). The **ANALYZE** commands produce on-screen readout of the constituent waves, whose amplitudes may be compared to find reflection and transmission factors at the set energy. Pressing [Ins] again restores the original wave.

The **EXTERNAL** entry in the **FEATURES MENU** assists the user in creating a xxxx.DAT file representing a user-defined potential U(x). The **EXTERNAL** procedure is especially convenient if U(x) can be represented over [a, b] by a few simple piecewise-analytic forms. Otherwise, records should be written to the user file outside the **EIGEN** environment. The xxxx.DAT file is accessed when the **EXTERNAL** option is selected from the **ENVIRON MENU.**

****

EIGEN3.EXE

HIGHLIGHTS

EIGEN3.EXE solves the Schrodinger *radial wave equation* derived for central forces. The resulting allowed energies E and *pseudo-wavefunctions* g(r) are displayed for quantitative on-screen examination.

Seven choices for the potential energy U(r) --- including a user-defined option -- -provide adequate diversity. Still more flexibility derives from the use of adjustable parameters and generalized units.

The *eigenvalues* E are found by an easy-to-use, interactive trial and error procedure, culminating in a visual display of the associated *eigenfunctions* g(r). Values for g(r) may be read directly from the screen, allowing full quantitative investigation of the waveform.

SPECIFICS

EIGEN3 solves the equation

$$\{- d^2/dr^2 + U(r) + \ell(\ell + 1)/r^2\}\ g(r) = E \cdot g(r)$$

for the pseudo-wavefunction g(r) = r R(r) expressed in natural units of distance and energy. (See the **EIGEN** synopsis for a discussion of natural units.) The bohr and rydberg constitute natural length and energy units for many three-dimensional problems.

Selections for U(r) are made from the **ENVIRON MENU**:

Spherical Well-	$U(r) = -U$ on [0,1] $= 0$ elsewhere
Quantum Defect Well-	$U(r) = -2\,[1 + \beta/r]/r$
Screened Coulomb Well-	$U(r) = -2 \cdot Z\, e^{(-r/\alpha)}/r$
Isotropic Oscillator-	$U(r) = (r - b)^2$
Morse Oscillator-	$U(r) = U\{1 - \exp[-(r-b)/(U)^{1/2}]\}^2 - U$
Charged Shell-	$U(r) = -U$ on [0,1] $= \Phi/r$ on $[1,\infty]$
External-	U(r) is user-defined

Entries for U(r) appear here in natural units; these units may be reviewed during execution by pressing [F2].

At the **POTENTIAL DISPLAY**, the **ANALYZE** function produces on-screen readout of the *effective potential* $U_{eff}(r) = U(r) + \ell(\ell + 1)/r^2$ anywhere in the **VIEW RANGE** [0, R]. Pressing Alt+ [V] overrides the default view limit and invites user input for R. Pressing Alt+ [U] initiates screen prompts for new values of parameters appearing in $U_{eff}(r)$, beginning with the orbital quantum number ℓ(default = 0). All changes are registered on the graphics display by pressing Ctrl+ [Enter].

A first guess for E is set at the **POTENTIAL DISPLAY** using the [Up]/[Dn] keys. Pressing Ctrl+ [Enter] initiates computation and generates graphics output for g(r) at the **WAVEFUNCTION DISPLAY**. g(r) is examined for suitability and E adjusted as necessary until no discontinuity is evident (as described in the synopsis for **EIGEN**). The **ANALYZE** function is used to study the resulting waveform. The **ZOOM** and **PAN** commands from the **FEATURES MENU** facilitate inspection of waveforms exceeding the vertical limits of the graphics viewport.

************* SPECIAL FEATURE **************

The **EXTERNAL** entry in the **FEATURES MENU** assists the user in creating a file xxxx.DAT representing a user-defined potential U(r). This file is accessed when the **EXTERNAL** option is selected from the **ENVIRON MENU**. For more on the **EXTERNAL** command, see the synopsis for **EIGEN**.

****

ORBS.EXE

HIGHLIGHTS

ORBS.EXE displays probability distributions $|\psi(\mathbf{r})|^2$ (the electron "cloud") for several of the lowest energy states of hydrogenic atoms. These are *density plots* , with screen brightness at any point proportional to the probability of finding the electron there. The display is confined to a plane defined by the z axis and a line normal to it (such as the x-z plane, or the y-z plane). Since the z axis is a symmetry axis for $|\psi_{n\ell m_\ell}|^2$, the density plot is identical in all such planes and the full 3-d "cloud" may be visualized by rotating the screen plot 360° about the vertical.

The view field is adjustable, thereby furnishing a global perspective or revealing finer structure as desired. The probability density $|\psi|^2$, can be read directly from the screen anywhere within the view field.

Plotting speed is increased at the expense of lower picture quality. The user is offered the choice of low, medium, or high resolution graphics.

SPECIFICS

Probabilities are displayed for the hydrogenic wave specified in response to input prompts for the shell number (n), the subshell label (ℓ), and the orbital index (m_ℓ). The **ORBITAL TYPE** is recorded on-screen as the leading entry in the **SETTINGS MENU.** This menu is filled out by prompts in succession for the **ATOMIC NUMBER** (Z), the **VIEW SIZE**, and the display **RESOLUTION.** Movement within the completed **SETTINGS MENU** is accomplished with the [Up] or [Dn] direction keys; changes to the recorded values are made by highlighting the appropriate entry and pressing [Enter].

Pressing Ctrl+ [Enter] opens the **VIEW FIELD** for plotting. The **VIEW FIELD** is square with the coordinate origin at its center. The width of this field (in bohrs) is the **VIEW SIZE**. Plotting begins automatically and continues to the limits of the **VIEW FIELD**, but may be halted at any time by pressing [End]. Values for $|\psi|^2$ anywhere inside the **VIEW FIELD** can be read from the screen using the **ANALYZE** function. The **ZOOM** commands listed in the **FEATURES MENU** enlarge or reduce the **VIEW FIELD** without having to re-enter the **SETTINGS MENU**. Return to the **SETTINGS MENU** for other changes by pressing Ctrl+ [Enter].

Note: The **CHANGE MENU** is superceded in **ORBS** by the **SETTINGS MENU**, which remains on-screen throughout program execution.